BIAS IS BEAUTIFUL,
OR
SWAN SONG
FOR COMMON SENSE

Mahmut Tolon

Cover design by Firdevs Ergen

The Need for Each Other. A picture of a baby hippo and a giant tortoise after the great tsunami in Asia on December 26, 2004. Pictures like this one and pictures of cats and dogs and swans falling in love with boats are seen frequently, as Konrad Lorenz, who received the Nobel Prize for his work on pecking orders in 1953, has demonstrated.

Source for back-over photograph John Roach for National Geographic News, January 5, 2006: http://news.nationalgeographic.com/new/2006/01/0105 060105 hippo tortoise.html

ISBN: 978-1-4196-6901-9

LCCN: 2007903775

To order additional copies, please contact:
BookSurge Publishing
www.booksurge.com
1-866-308-6235
orders@booksurge.com

Printed in the United States of America

Table of Contents

Figures

Introduction

In our world today, a scarcity of fossil fuels and drinking water is becoming increasingly evident. "Global warming" is now part of our everyday vocabulary. People are living longer with fewer young people to support them. Job creation has not kept pace with population growth. Because of technological developments, more and more people are jobless. Violence is our next-door neighbor, wherever we live. We are at war with our planet and with ourselves. The twenty-first century seems to be a century where haves and have-nots, young and old, and humans and the rest of the earth will have to find ways to coexist peacefully.

The key to solving most of our problems is to change our approach to these problems. It is a fact that we are all, by nature, biased. People tend to think of themselves as either the center of the world or part of the herd. The roots of bias lie within our biology and in the distinction between self and non-self. Equipped with a clear understanding of bias, we can begin making a decisive change. Bias, which forms the basis of our everyday existence amid the "rat-race," when understood better by more people, could cease to be a threat and almost become a cause for entertainment. The time has come for us to mold our own destiny.

In the following, I will try to restate the story of human population and culture, briefly touching on international relations as well as personal experiences (probably not so different from the ones you have experienced) in an attempt to show the similarities between the two and how bias has affected us in history, how it has helped to form the world's cultures, and how it still affects our everyday lives. When the human population was under 2,5 billion, Albert Einstein said that it is easier to split an atom than to break a prejudice. This book takes prejudices apart. Bias can if understood widely, make a huge difference and life a lot more fun.

About the Author

Mahmut Tolon was born in Istanbul in 1950. He attended high school and studied medicine in Kiel and Bonn, Germany. During his studies he worked in Bath, England, and later under the DAAD (German Academic Exchange Program) in Australia, at the Sydney Children's Hospital, and in Broken Hill with the Flying Doctor program, as well as at the psychiatric hospital in Provo, Utah.

Dr. Tolon currently lives in the small town of Urla (site of the prehistoric harbor Limantepe) overlooking the Aegean islands and teaches a postgraduate course on "Longevity and the Co-existence of Cultures" at the nearby Dokuzeylul University in Izmir.

PART ONE
BEGINNINGS AND BASICS

~1~

A Bomb

Night shift had begun in the medical clinics of the University of Bonn in the park in Venusberg, Germany. It was a mild winter evening, and a light shone from a small room on the second floor of the Hygiene, Microbiology, and Immunology Building. I had stayed late to finish some experiments, and I smoked a pipe as I waited for 11 p.m., the time when I could begin counting the microbe colonies.

At the Brain Surgery Clinic, I collected samples of air during surgeries to determine how many airborne germs were being killed with the ultraviolet (UV) rays used for this purpose. I enjoyed this work, which was the subject of my doctoral dissertation. The institute had become my second home, and my professor had given me a job as a research assistant and my own office.

While waiting, I picked up the book I had purchased that afternoon and started reading: "As you read these words five humans, mostly children, will die of hunger. At the same time, forty babies will be born." I gave it some thought: If I read very slowly . . . if it took me one minute to read those words, in an hour three hundred people would be dead and twenty-four hundred would be born. But if twenty-four hundred were born and only three hundred died, what happened to the rest? They lived, of course! But how would they live? How could they ever find good schools, qualified teachers, or even enough food, for that matter?

I continued reading with growing interest, finishing the book in three days despite my busy research schedule. The book had a tremendous impact on me, and some of the terms and ideas in it lodged themselves in my brain.

The day after finishing the book, I excitedly called my father. I wanted to obtain the publishing rights to this book, titled *The Population Bomb*, by Paul Ehrlich, so that it could be published in Turkey. My father said, "For heaven's sake, son, forget about these things. Don't get mixed up in politics. If you publish such a book, you'll immediately be labeled a communist or something. You're going to have a good life as a doctor, so don't mess it up."

But I was convinced of the importance of the book, and some further nudging convinced my father as well. He was almost seventy years old and was an ex-member of the Conservative Party in the Turkish Parliament. Few would dare call him a communist in Turkey, so the decision was made to use my father's name as the editor. I wrote a letter to the author and the publisher: "We are convinced of this book's importance and want to publish it in our country. However, we do not believe it will earn much money, so . . . we would like to suggest a gentlemen's agreement, that if money is made, we would like to offer to share the profits."

Barbara Ballantine, representing the original publisher (Ballantine), agreed to our request. I energetically began translating the work into Turkish. A German version of the book was purchased for my father, who spoke little English, and together we produced a translation. Unfortunately, we were unable to find a publisher in Turkey. Nevertheless, I had the book printed at my own expense, even before the publication of my dissertation on the effects of UV irradiation on airborne bacteria.

Over thirty years have passed, and I am still interested in the topic. I have realized how big the issue of world population is and believe that it may be even more important than any other issue facing society.

As a doctor, I may have been able to save the lives of quite a few people throughout my career. And more lives have probably been saved as a result of the research I've been involved in. But what if this number of lives saved was to increase tenfold or even one hundredfold? Would it really matter?

Things change. The way we see things as a helpless child is different than how we see things as a young man or woman, and the perspective of an elderly person is different still. The healthy person can only show sympathy but cannot understand the sick. The rich have a different view than the poor. If we follow this line of thought, we end up saying no two men are alike in their feelings, which also change with time. In other words, everyone's view of things is biased by his or her circumstances, by climate, by age, by health, and by wealth. So can bias be a common denominator? While I am gazing at the islands on the Aegean Sea and writing these per-definition biased lines, the copulation—Oops! Sorry!—I mean, of course, the population of the world increases by approximately seventy million every year. That means an addition of double Canada's and Australia's entire populations combined year after year. But don't worry: This book will not talk about population forever. There will be longevity and lots of sex in it. And it will definitely have a happy ending.

The following will be easy to follow if you are interested in biology. I will try to outline an understanding of the world in a hasty (but I hope entertaining) tour. Readers are encouraged to look up a term or two or, if they want, to check a fact. At the end of the book, you will also find notes about further reading for each part.

There is a solution to the world's problems, and that is either to freeze or even decrease world population.

The United Nations has a beautiful slogan: "Everyone counts." This statement is absolutely correct. Anyone is or has the capacity to be as qualified as any other person, at least as far as the population is concerned. In fact, you should read this book as a fellow teacher reads a colleague's written words, takes notes, evaluates, and then checks his or her colleague's conclusions.

Because the story of humans is the subject of this book, let's have a brief overview of demography, or the study of human population. Imagine! You are all alone in a forest. You will need food and security. You will seek out others like yourself to help you to survive. When a group first gets together, its first priority is survival. Its members will need to communicate with each other. This is the basis of culture.

As long as people feel safe and have enough food, they will freely copulate. This is the basis of overpopulation. Because we all are selfish and have different priorities and interests, as the members of the group increase, the pecking order (the foundation for any hierarchy) will create conflicts. This is the basis for our bias. We will naturally split off into different groups, or herds, and compete for better pastures or hunting grounds. Our codes of communication or languages will begin to differentiate, and new leaders will emerge who think that they are very special. Different cultures will develop. Out of this will also come group identity and prejudice against other groups.

As cultures develop, people will tell each other stories about catastrophes and how to understand them, deal with them, and survive. They will develop moral codes that will evolve into belief systems, offering a happy ending to the trials and tribulations of daily life. These belief systems will turn into religions whose dictates will be passed on to people's children and their children's children, and these children will believe that their religion has enabled them to survive and succeed.

If you think about human development and the efforts required to educate each individual, it is not hard to understand why people identify with their belief systems, habits, and culture and want to pass these on to their offspring. Each pupil will have to sift through the information provided for him and categorize it as "useful" or "not useful," as concerning the self or not concerning the self. But he will have to trust in the validity of the information and believe in it. Although in this book I am quoting from external sources, I have to believe in what I am telling you. There is not enough time within our lifetimes for me or the reader to check all this information. I believe in what I am saying because I trust in the integrity of the scientists and thinkers who have worked in their respective fields. I have worked in different fields of science myself, and I have seen many scientists spending decades checking and rechecking their results. I have also seen some of them tell lies, or omit part of the truth. Scientists work within the limitations of human behavior, after all, and are susceptible to greed, envy, stupidity, overconfidence, and self-interest. Even so, the vast majority of them still try to tell the truth. When the scientific community agrees amongst itself, after all the checking and rechecking, then we can assume that the data which emerges at least approaches the truth.

Demographers estimate the population figures on the basis of archaeological sites and their capacity. Written demographical data are known since about two thousand years ago and have become very precise in the last two hundred years. Let's glance at figure one:

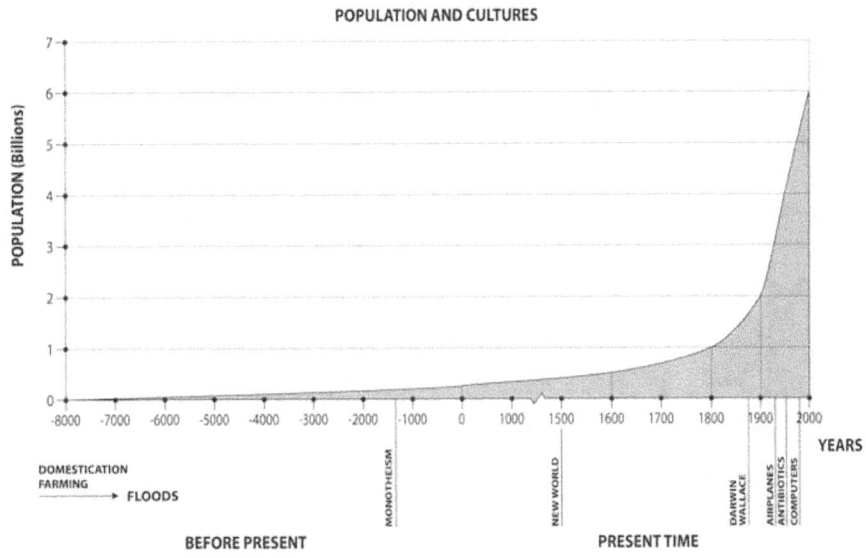

Fig. 1: The Population of the World: From the Birth of Jesus to the Year 2000

Simple, isn't it? The graph shows a slow and then increasingly rapid growth in human population. The world had about 250 million people when Jesus was born, 500 million when America was discovered, and more than six billion in the year 2000.

Let us now look at the reindeer. In the year 1944, twenty-nine reindeer were released on the three-hundred-square-kilometer St. Matthew Island in Alaska. These reindeer had a peaceful life with no predators. They multiplied at such a rate that there were more than six thousand of them by the year 1963.

By then, the island had begun to become barren because of overgrazing. Vegetation was unable to replenish itself properly. As the population of reindeer increased, food became insufficient. When the harsh winter of 1964 set in, the deer perished. Piles of reindeer skeletons littered the landscape. There were only forty-one females and one sterile male reindeer left; the reindeer were not able to survive as a species on the island anymore, and the population died off one by one.

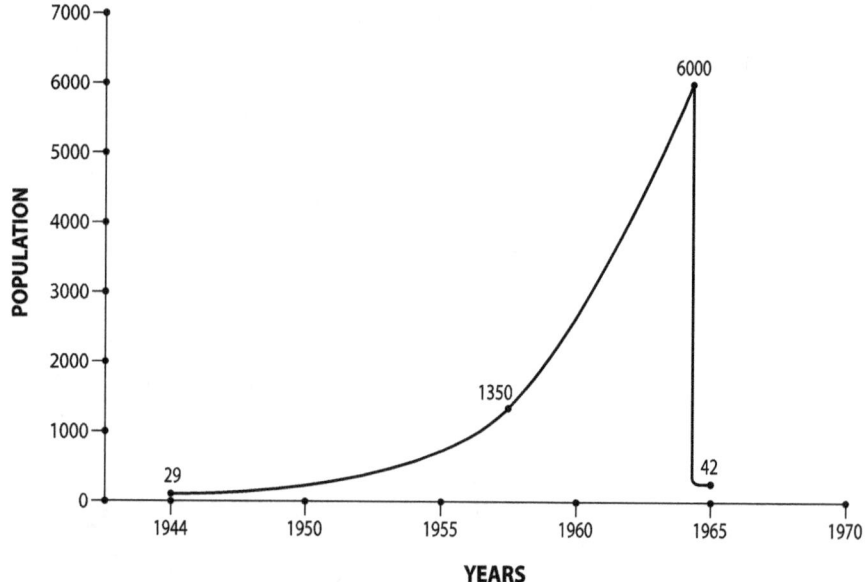

Fig. 2: Population growth of the St. Matthew Island reindeer herd. Actual counts are indicated on the population curve.

Source: David R. Klein, Alaska Cooperative Wildlife Research Unit, University of Alaska College: http://www.populationaction.org/resources/publications/naturesplace/NaturesPlace.pdf; http://dieoff.org/page80.htm

There are no hidden ghosts or devils; this is a straightforward experiment in nature. It is easy to understand and your opinion is just as important as that of celebrated scientists who have also grasped these simple facts. As a farmer who has practiced medicine all his life, I will try to share with you what I understand from biology, which, I believe, is more captivating than any crime novel.

Bias Is Beautiful

One can make a comparison between the population of reindeer, which is six thousand, and the population of human beings, which is more than six billion. Basically, be it the reindeer herd on the island or the ants in an anthill or the microbes that grow on a sea moss called agar agar in a petri dish, the result is the same: When a population reaches a point at which nature cannot support the number of individuals in the colony, the colony begins to perish. The predictions of biology are no different for human beings. The philosopher Ludwig Wittgenstein's teacher and friend, Bertrand Russell, said, "Prediction is one of the duties of science."

The reindeer island is a metaphor (microcosm) for the earth. The earth becomes barren due to "overgrazing." If not enough food is available, the population becomes undernourished and more susceptible to environmental impacts and diseases, but also nature starts to go off balance. This process is a trifle more complicated on a larger scale, because extra food for humans can be produced in greenhouses and with modern agricultural methods, such as drip irrigation. But the imbalance with nature is still there. With fewer plants in the environment, less oxygen will be produced and more carbon dioxide will be released into the atmosphere. We have started our "overgrazing" and are heading toward a shortage of air and water. The fight to save the rainforests is not the concern of a few overanxious environmentalists, but a central theme of our survival.

~2~
Bomb within a Bomb

Let us continue with the story of *The Population Bomb*. I paid for the printing of six thousand copies of the translation of this book. About four thousand copies were distributed, sold, or given away. A few newspaper columnists wrote articles about the importance of the subject and the book. The leftover copies were stored in my father's flat in Ankara, Turkey. About ten years later, in 1985, I placed most of the remaining books in the mailboxes of parliamentarians in the impressive parliament building in the capital.

These books were disposed of and entrusted to people who I believed could make a difference. In the meantime, it was becoming more and more obvious that the population had increased dramatically, and was exploding to dangerous levels.

When the population of the world reached 2.5 billion in the 1950s, scientists began to issue alarming warnings. One of the pioneers of modern demography was Alfred Sauvy, who coined the term "third world" in 1952 to refer to the poor countries that had the right to a larger share of the world's riches. He called the West the "first world" and the communist empire the "second world." For my own purposes, I have modified this expression and believe that there are hopeless, jobless, and uneducated people within all countries, varying in percentage but similar in many ways. This "third world" is the main target of the actions that must be taken to ensure all of our survival.

I understand the "first world" within every nation as the self-sufficient citizens who contribute to the common good by acting as producers. I consider the "second world" within every nation to be the government employees and uniformed and nonuniformed bureaucrats. Of course, this is only my own biased working hypothesis, no more. Rather than speaking of the "third world" as "a few poor countries in need of aid,"

I believe there is a "third world" of hopeless, aimless, and uneducated people in every country.

Thankfully, reindeer also live on other islands and continents. The island called Earth is fairly big, so there's no need to be pessimistic. We can continue to enjoy life. The solutions are clear and simple, and the more people who understand the dangers of overpopulation, the faster we'll be able to reach a solution. The world has become a small island thanks to the unbelievable level of communication we now enjoy. The days of Pony Express riders carrying letters on horseback are long gone. We get information instantaneously. The information is still biased according to the interests and the background of those who distribute it, but we have more unbiased information than ever before—and from a multitude of sources.

We know now how the animal called "man" reproduces, but there is within this increase another bomb, which I shall try to convey. Society's realization of the new population bomb began in the 1990s. We probably need another ten years before we can say that the whole world has understood the seriousness of the new problem: the Age Bomb. One of the pioneers in the field of the aging world population, Peter G. Peterson, stated: "Two-thirds of all people who have ever lived to the age of sixty-five are alive today." This sentence is so striking that it needs further analysis. I had to repeat it to myself five or six times in order to grasp it.

Mummies and ancient bodies found in ice tell us that the average life-span in the time of the Egyptians and ancient Greeks was only thirty years. The oldest citizens reached the age of fifty. The ancient Egyptian Pharaoh Ramses II was a recorded exception, living past eighty. Not until the first millennium CE (common era or present time) do we start to encounter plausible records of more people reaching the age of eighty or more.

In the Middle Ages, only three or four people out of a hundred were able to reach the age of sixty-five and above. This ratio increases progressively in the 1900s. The situation is more striking now: One out of two children born in developed countries will live more than one hundred years.

The life expectancy of humans increased by three months each year in the last decades of the twentieth century (and a little more for women). When Otto von Bismarck passed the first retirement pension bill in Germany in 1889, the average lifespan was forty-two years, so that only 4 percent of the population benefited from having paid into the system. In other words, a hundred years ago, the viability of pension funds was not much of an issue, because people conveniently died before retirement age.

Bismarck was a successful example of a politician who heralded extraordinary new directions. Aside from introducing retirement pension laws, he galvanized the population around the concept of a united Germany. He also obtained considerable concessions from the church. Human history is like a constant tug of war between belief systems (and their desire for moral control) and the individuals or the silent majority who seek to hold onto independence of thought.

Since translating *The Population Bomb*, I have been interested in the subject of human population. Whenever I, as a physician, have tried to discuss the population problem with the politicians or generals I was treating, they would start looking at their watches and their eyes would glaze over after about fifteen seconds. One can get rather close to powerful people while treating them, though they don't even come to the doctor's office alone. When I've brought up the subject, they've turned to their consultants or assistants and announced that it was a worthwhile subject to investigate in depth. Of course later, it is impossible to reach them.

Still, politicians (like other quick-witted people) occasionally get it. But on the urgency scale, their agendas prevail. They also only have twenty-four hours every day and one life to live. And it's perfectly normal that they are not interested in hearing the story of "reindeer island" at the doctor's office. As with every individual, a decision-maker has both enlightened and dim moments. Requests for favors ("Help us, oh exalted one") can wear out any leader. Consequently, they surround themselves with "yes men" with compliant dispositions. Able advisors contribute to the system and do not cause problems; on the contrary, they solve problems and show loyalty to their leaders. If the leader dies or retires, the assistant who takes over, because he is less creative, usually rules by surrounding himself with people who advocate the status quo and do not make waves. When a society has "democracy," change can take place at a more leisurely pace through elections, whereas in aristocracies, changes happen either at a faster pace or not at all. When there is a democratic election, we are usually spared being saddled with an unfit ruler who inherits the position because of blood ties. But then, of course, a populist leader has more of a chance to be elected.

~3~

Of Politicians and Cabbages and Kings

No matter how honest and educated, a man in a position of power develops a feeling of superiority. It is natural that this feeling then leads to "professional deformation." That is why it is a wise policy to require such people to take a sabbatical every four, five, or seven years to turn their attention elsewhere for a while. Sabbaticals aren't only good for politicians; they are also of utmost importance for doctors, lawyers, judges, prosecutors, and policemen, as professional deformation or formation occurs in all these callings. In fact, any routine causes an adaptation, hence a formation and deformation. The truth is, we are by nature fond of routines; we are almost addicted to them, because they give us a sense of security. This is neither good nor bad; it simply is evident and in many cases something that is desired. Think of the marathon runner training for a race. After a few years of training, his capability to run will change, as will his understanding of life.

The human being lives into the future, but by nature understands only the past. We "naturally" react to daily routine with the sum of our experiences. Societies or tribes react with whatever they have in their collective national memories and values. For example, it took the Great Depression in the United States and the Weimar Inflation in Germany to reach an economic understanding of the impoverishing effects of inflation. Economically, it takes a society a few years to react to a new situation, "a braking time." People tend to go on with their spending habits for years before they can adjust to a new economic reality, such as the present oil shortage. The same is true for a human being; if the "hunger" threshold is wrongly set in his mind, then he realizes that he is full only after overeating. Overeating calls for more insulin in his blood, which in turn gives him the urge to eat again in a short while. The insulin swing causes the blood sugar level to drop, which leads to the false feeling of hunger and the subsequent inflation of fat cells. As a person gets older, experience, intelligence, and professional intervention can help him to correct this tendency. This principle of needing expert advice also applies to societies with economic inflation. For that matter, expert advice is also often needed to mediate the disputes of two neighbors or tribes.

In the last couple of years, as a retiree freed from daily routine, I've had the opportunity to reread books on birds and bees and cabbages and kings. Since Bismarck's time, old age has been defined as sixty-five. Actually, at age thirty-five, the body begins to give its first indications of aging. At forty-five, the eyes can't see as well as they used to. And the deterioration is not because "you read too much." Whether you read or not, the eyesight starts deteriorating around age forty-five.

However, with modern healing procedures, solutions are easily available. Sixty-five is no longer an important milestone. Eighty or eighty-five is more the age that deserves the term "elderly." And those who reach that age in a relatively healthy state set their sails for one hundred. Sixty-five-year-olds climb mountains, surf, or ski if they are so inclined. Nowadays, in developed countries, it is necessary to move the old-age frontier to eighty-five. Anybody under forty thinks sixty-five is really old. Anybody in their fifties thinks of people in good health and active in their eighties as "old." Is bias not beautiful? I'd like to impress upon you the size of the aging population by making an educated guess of my own: since time began, perhaps 85 percent of the humans who reached the age of eighty-five are alive today.

If the overaging population is increasing at such a rapid pace, who is going to pay for their retirement? Let's leave that subject to politicians.

Look at a map of the world, paying particular attention to the countries in Africa and Asia whose borders were drawn with a yardstick. Many politicians have little idea that many nations' borders were drawn, often ineptly, by colonial powers. The war to "bring democracy to Iraq" shows us how little they understand about inept borders, in this case, those drawn up by the British. Present-day Iraq has a number of different tribes (who actively dislike each other!) living in the same geographical entity. And they have no idea that this happened after World War I. Many think Iraq or Rwanda or the former Yugoslavia are "nations" that just need the one-man/one-vote principle forced on them—and that then there will be no more wars.

It seems we haven't got the luxury of leaving problematic issues to politicians; we must address them ourselves. Remember? Everyone counts.

After glancing at these maps, we wonder why politicians and consultants, mostly after World War I, drew these borders so ineptly. Why? Because they didn't know much about biology! Borders were drawn at a conference table—and usually without the presence of any natives from the places in question.

The mention of World War I takes me back to my school days: I didn't particularly care for history courses. Usually it was about what King

Henry this or Louis that or what Emperor Walrus did or wrote. I became instantly sleepy and took off to a favorite and widespread occupation: daydreaming. The only one of the Henrys I remember is Henry VIII—he lived in an important period for religion and had six wives who could easily be remembered with "divorced, killed, died, divorced, killed, and survived."

When Jesus was born, there were perhaps 250 million people living in the world. The city of Ephesus (now a tourist attraction as an open-air museum with beautiful ancient marble alleys in western Turkey) was once the New York or Beijing of its time, with about two hundred thousand people living in it. Why was Ephesus abandoned? What happened that the pyramids were forgotten in the desert (except by the thieves) until they were "re-discovered" over two thousand years later.

What factors caused Hattusha, the capital of the Hittites, to become a ruin in the middle of a stony steppe? Hittites, who also lived in Anatolia over a thousand years before the ancient Greeks, were the contemporaries of the old Egyptian Pharaohs. They fought the battle of Kadesh with the Egyptians some 3,300 years ago, and the truce after the battle is the first-known written peace treaty. Both sides claimed to be victorious after the wars, of course, writing their side of the story. Bias is actually a part of history and we are not as different from our ancestors as we think. The Hittites, by the way, had an interesting name for water. They had the same name for the precious drink some four thousand years ago as the Anglo-Saxons of today. They called it "water"!

History is written by the victors and this makes it rather lopsided. The Greek meaning of the word "history" is "inquiry." History was for a long time a simple repetition of biased "facts"—one historian copying from another. Like anthropologists inquiring about the meaning of bones, which are so-and-so-many thousand years old, modern historians are starting to search for the meaning of what has happened.

When historians speak of World War I, the fifteen million people who died in the war are commemorated. When I look at the same time frame from a biological standpoint (the world population at that time was just over two billion), I think of the Spanish Flu (known as "The Great Influenza Epidemic") that took place at the same time as World War I and killed more than sixty million people. Both warring sides' press was so heavily censored that scientists only got at the truth about the scope of the epidemic in the 1980s.

We're fast reaching an enlightened age. It was forbidden to speak of the Spanish Flu during World War I. Both sides hid their losses because they thought the other side was using a "biological weapon." Yet the death toll was neither the wrath of God nor the "other side's craftiness." It was an epidemic caused by a simple virus consisting of genetic material and a fatty string. And it spread rapidly in the overcrowded and unhygienic conditions of trench warfare.

But all this has been left behind. Technology is advancing quickly. In the nineteen hundreds, people benefited from electrical energy, then the telephone; first the telegraph and fax and now the cell phone and Internet have entered peoples' lives. Of course, we must not forget radio and television, cars, airplanes, refrigerators, washing machines, and microwave ovens. All these changes have occurred within the last hundred years, and not just for the use of a few. Their use is widespread.

When people first became aware of the population explosion, they worried there would not be enough food to feed everyone. Developments have shown that producing enough food is not the problem, but distributing it fairly is. The problem of hunger is actually one of education, distribution, and communication. Those who can help in the war against ignorance and the slow pace of development live in large cities, in security-conscious residences. They don't want to be in dangerous, crime-infested neighborhoods. That makes sense, doesn't it? Birds of a feather flock together; it is an instinct even older than humanity.

However, that means that the poor and uneducated get more cast out. The well-to-do, with their own values and needs for security, want to stay with their own kind. Political systems that did not wish to acknowledge this reality were formed and fell. Biology does not work by coercion. So people won't let themselves be forced into doing what they don't believe is in their own interests. If they have no choice, people might put up with a system they don't believe in for a short while, but then everything will return to its own nature. So, if the main reason for the bad distribution is communication, it stands to reason that the most successful people in today's world ought to be those involved in communication and those making discoveries in that area; and isn't that so?

Computer producers Michael Dell and Steve Jobs are two examples of the thousands of people who became wealthy through technology. There might have been monetary differences among individuals in that field, but they earned their money through creativity. These successful people are still relatively young and actively working toward ameliorating the world's problems through the various charitable foundations they have endowed. When these people were born, the world population was around 2.5 billion. And there are more than 6.5 billion of us today!

If we look at humans as a biological herd (and it would be unreal to look at them in any other way), the more successful live longer and are more influential. Successful people are in a better position to solve the world's problems. Simply stated, because they live much longer, they are able to bring about ideas and communication reforms faster and more actively than governments. You can be sure that a person who matured and earned his money under free-market conditions is not going to spend his money with an "employee mentality." And the effects of the foundations these successful individuals finance will be increasingly felt.

It would seem there is no reason to be pessimistic. And I am not being ironic. Nor am I being cynical! There are a lot of reasons to be optimistic in today's world. The problems we face are not insurmountable, and we can all rally around the plans that need to be put into effect.

~4~
Hardware & Software

When we look at what we know, what needs to be done will become obvious. Be it good or bad, we have some patterns of behavior that are in our "hardware" and some that are in our "software." Not all of what we have in our hardware is good; in fact, when we look at history, we may not be a very intelligent species. An example of our behavior that is part of our hardware is our pecking order. Our best-known instinct, also in our hardware, is the drive to multiply and hence to survive. Training for our place in the pecking order is a sport where winner takes all. This is also part of our survival drive. In biology, often enough, only one can survive.

Another behavior deeply imbedded in our hardware is our tendency toward addiction, or in more positive terms, our affinity toward routine. A simple example is our fondness for sweets. As the human evolved in the rainforests of Africa, which later became the Savannahs, sugar was rare, and later, during the ice ages, our ancestors were (and hence we are, by nature) programmed to eat as much sugar as possible whenever it could be found. Sugar is an important factor influencing fertility. When the female body does not get enough of it during shortages, the female estrus cycle changes and does not allow pregnancy. As the female can't care for her offspring properly in times of crisis, this relationship between sugar and fertility is a built-in natural control mechanism. We can also observe the relationship between sugar depletion and lack of fertility in extremely thin female marathon runners.

After the victory of our ancestors over nature—and mainly after the Industrial Revolution—more and more sugar (or what we today have evolved to call "junk food") was available to the masses. This dietary change, along with advances in medicine (mainly hygiene and vaccinations), were the main causes for the population explosion. Most of us now know what an insulin swing is and realize that this is what makes us seem to be hungry again shortly after a big meal. Although our "intelligence and knowledge" tells us that it is not good to eat again, the flesh is weak.

The readiness to travel is in our hardware as well, and not only the "walkabout" of the aboriginals or "wanderlust" of the Germans, but also the general tendency of modern man to travel whenever he can afford it. We might refer to this as "tourism" or "weekend trips." Driving long distances for a few hours of entertainment or food intake gives us a feeling of accomplishment and distances us from our daily worries.

Bias is another behavior that is in our hardware. When we understand and define our hardware clearly, beliefs in complicated theories become obsolete:

For example, Sigmund Freud's ideas partly became obsolete because of advances in science; we now know, to put it simply, that love is partly oxitocin, that the flight-or-fight reflex is adrenalin, and that sleep and happiness are partly opiates (endorphins) and partly melatonin circling in our blood. This by no means suggests that Freud has become unimportant. He simply didn't know of these hormones and chemical substances that regulate us. No one did. The emotional molecules were not a part of human culture then. He still is a pioneer and a prophet of psychiatry, but I believe within a few more decades his teachings will be of interest mainly because of their historical value.

Another example of biological hardware clarifying theory is the stomach ulcer: In the last century, the bestselling drugs were products to counteract stomach acid, believed to be the cause of ulcers. Now it is understood that ulcers are caused by bacteria in the stomach. In 2005, Australian scientists were awarded the Nobel Prize for their work on helicobacter pylori. This example is not unique; many branches of science have undergone a revolution in the last ten years.

But hardware, which we all have in common, and "software," which

Bias Is Beautiful

is the product of our neocortex and a recent development in our evolution, see and act hand in hand. The unique "software" finally is the sum of each person's experiences and education. Psychologists and neurologists tell us that no two humans are alike. Different types of feedback cause different neuronal synapses and everybody ends up being unique. Toxicologists will tell us that no two humans react alike to different chemicals. Immunologists and histologists also agree: No two people, not even twins, are alike! In short, we are all different, but every one of us has the potential to be very important and could bring out the decisive different approach for us all. So everyone counts.

Two groups of scientists have shaped our current understanding of the universe. Because their findings were not based upon religious or patriotic principles, which one simply accepts or rejects as a matter of faith, belief, or belonging, their work was seen as a threat to their societies.

The first group of scientists consists of Avicenna, Bruno, and Galileo, and somewhat later of Newton and Einstein. They explored the interrelationships of physical properties and of the planets. The second group consists of biologists who revealed how animals (including humans), plants, and physical environments influence one another and how change in one system necessarily generates change in the other systems. Charles Darwin is probably the most important, because his work helps us to understand the crucial role the physical environment plays in shaping human culture and behavior and how we can thus shape the future, rather than being simple pawns in what may happen to us. Though we shouldn't try to label the most important scientist, many agree that Darwin's contribution to humanity was the most important. However, even the most revolutionary invention or theory is only a product of our mutual common sense; if scientist A had not described it for us, scientist B would have!

Of perhaps equal importance, though, are the others who built upon these and other scientific ideas through their inventions and organizational efforts. Some of the thousands in this category are Thomas

Edison, Ford, and Bill Gates. These individuals are household names today, people who through hard work and by using their wit and knowledge were able to tap financial resources to invent and market useful products. In the process, they created many jobs and changed the way we communicate with one another and deal with everyday problems.

~5~
Key to Life.
How Do We Know?

In daily language, the word "theory" is used as a synonym for an idea or hypothesis. The hypothesis for evolution has been checked and double-checked by the scientific community over 150 years. It has become easy to demonstrate and prove evolution to the biochemically, genetically, or microbiologically schooled person: All you need to do is put a colony of microbes under some sort of stress and watch the mutations and the change in the population, usually within a few days. The problem is only for the person with no such background: Even this proof seems to be metaphysical for him.

People who believe in an intelligent design will cite the "second law of thermodynamics" and say that it is impossible that such a superior being as man could have evolved if left on its own. As one gets curious and reads about this "second law of thermodynamics," one sees that it is about entropy and nothing else but what we know today as Murphy's Law: Everything that can go wrong will go wrong. This alone shows how biased the perspective is. If you start from the point of view that humans are perfect, then you will cling to such "laws of physics." If physicists or chemists like calling their common reactions "laws," then by the same measure we should all call evolution the "constitution." Yes, even the law of gravity is no exception. Take a handful of common sense and think for a moment what you could do with the law of gravity if there was no life. Evolution helps us to understand how we came to be and not why we are here, of course. It is true that scientists as a group, or herd, have veered off in wrong directions, notably in Nazi Germany, so now it is very important that everyone can voice his or her opinion as a corrective.

A basic understanding of the atmosphere and the oxygen we breathe, as well as the water that we drink and out of which we came, is essential for daily decisions regarding our survival. Allow me to take you on a brief journey into chemistry, space, and biochemistry, always with the concept of cultures, bias, and overpopulation in the back of our minds.

In a branch of geology called sedimentology, scientists get clues regarding the world's age by studying rock layers and formations. Biologists also have methods to make their own estimates, and so do physicists. Allowing for margins of error, when the results of such different methods agree, we have an accepted "fact." The other option is simply not knowing. We don't like saying "I don't know," so each culture has rules, theories, and belief systems that shape us.

Molecules have their own built-in rate of destruction. Nothing is eternal. Most matter contains radioactive components. These isotopes have a radioactive half-life, so scientists can measure the radioactivity and calculate the age of a bone or that of a stone rather precisely. When you read a date like eleven thousand to twelve thousand years ago in this book, the large margin is given because of the inadequacy of different data. Besides, we must consider differences in calculating techniques in the past or the fact that new discoveries are still being evaluated.

The radioactive isotope of carbon, with a half-life of 5,730 years, is used to determine the age of various artifacts. With this method, it is possible to measure sixty thousand years back. For example, uranium turns into lead with a half-life of 4.5 billion years. When we speak of half-life, it means that the radioactivity of carbon is reduced by 50 percent in 5,730 years. Willard Libby was awarded the Nobel Prize in chemistry in 1960 for his work on this topic, which helped revolution-ize our understanding of anthropology.

The farthest-traveling noise humans have created is the radio wave. We know its speed. We know that it takes hundreds of years for radio waves to reach the nearest star and that humans have only been engaged in listening for messages from space for the last few decades, so it isn't surprising that no contacts have yet been recorded.

By calculating the roundtrip distance of these waves, it is not difficult to conclude that for all practical purposes, we are alone in this universe. Realistically, we will continue to reside in this world without help from extraterrestrials for at least another few hundred years. This privacy should make us happy!

And what would have taken place if aliens who had also discovered radio waves had the technology to travel to distant planets such as earth? Would we enjoy their visits? What if when the little green men land, they treat us the way we treat other animals and plants? When we think of how humans were formed through trial and error, the possibility of an invasion by aliens is even less probable. We'd better leave that encounter to Hollywood for the next thousand years or so!

It isn't possible to digest the idea of infinity. Mankind is fascinated by fantasies of parallel universes and time travel, and these have become a staple of science fiction. Is there intelligent life on other planets? Theoretically, there is no reason for there not to be. However, realistical-ly, it is best if we spend the next couple of centuries addressing more mundane matters, assuming of course that we are able to avoid consum-ing our planet during that time. Other species will surely survive on this planet no matter what we do. The question is, will we?

~6~
Everything Is Relative, Even Time!

What have scientists learned since Einstein? That time is also subject to gravity and that there is no independent entity. Even time is relative! What is called "time" was once measured by sundials, and then in the 1950s it began to be measured by atomic clocks, almost microscopically. Time on a mountain goes faster than by the sea, and indeed, a man's feet age less than his head, though the difference is measured at a rate smaller than a millionth of a second. In astronomic dimensions, time lengthens and shortens between other planets and stars, depending on different rates of gravity. In other words, when it approaches planets or large masses, time bends, like a spider web, and slows down. It completely stops in black holes in the cosmos, where theoretically matter can not exist. But none of this applies to daily life!

Perhaps if we could travel to a black hole and stay a while, time would stand still. But there we would cease to exist. When you look at the sky on a dark night, you see stars. When one recalls that each of those stars is a solar system and that there are billions of such universes, one is truly astounded by the concept of infinity. Despite the infinite number of solar systems and the therefore infinite possibility of life elsewhere in this universe or another one, the chance of all of us moving there after we destroy our own planet is negligible!

Newton lived from 1643 to 1727 and is the man known for describing gravity. Currently, well-read people often espouse notions such as the following: "Newton's theories are no longer valid—quantum physics and Einstein's relativity theory have disproved them." This belief is only true if one-upmanship is the aim of such a statement. Of course, Newton's theories are correct and necessary for earthly physics and daily life. What is really interesting is that until that famous apple fell on Newton's head, no one else had noticed gravity.

Of course, gravity was noticed and applied in daily life. How could the pyramids have been built without a notion of gravity? Biruni(970-1048) also called as the first anthropologist, concluded in 1030 that if the world rotated, all stones and trees would fly off if it weren't for a force that pulls everything to the earth. But Newton is the first whose work survived into the present; he also described his findings in mathematics and Latin (both of which I define as difficult-to-understand languages, math having the benefit of being the least biased language, at least in the way I see it). Newton's work was a scientific landmark. You can rest assured, though, that gravity would have been described without Newton, relativity without Einstein, blood circulation without Harvey, and evolution without Darwin.

~7~

Looking at the Sky

When looking at the data in this book, we should not forget that for each calculation and each conclusion, lifelong measurements have been conducted. There were pushing and shoving and dirty tricks. Instruments were invented; firms were founded that manufactured these instruments. Firms went broke. While all this was going on, people married, had children, were fired from their jobs, divorced, fell in love again. Some unfortunate people were even burned at the stake. It seems miraculous that we have come as far as we have, yet our essence hasn't changed much in the two hundred thousand years since we first populated this planet. Every day, people all over the world are being killed for their ideas or their money on streets not far from us. The stories you read in newspapers and see on television are not fairy tales.

In the history of science, decision-makers—in the Middle Ages, these were generally men of the cloth—disturbed by new ideas, became fearful of losing their dominance and began exerting pressure. Galileo

(1564–1642), the scientist who died in the year of Newton's birth, is known as the person who argued that the earth was a moving body revolving around the sun. He was tried by the Inquisition and made to repudiate his beliefs. Fewer know that years earlier, in 1600, Giordano Bruno, who was educated as a man of the church, was burned at the stake for similar beliefs, which he refused to renounce. Maybe it is typical for human nature to remember and champion the scientist who bends and stays alive rather than the one who refuses to give in and dies!

Religions were organized to resist prevailing authority, but soon became the institutions exerting pressure on society, and they also began fighting with each other. Religion is about power and control? Not about love and peace? Oh my! Not the true word of God but rather a human invention to fulfill the desire for control?! Still, we must not overlook the fact that religious orders educated the likes of Al Biruni, Bruno, and Darwin.

Throughout history, we see that when a fact could be proven scientifically, religions, after showing initial resistance, reached a truce with those ideas and became institutions that learned to live with reality. Philosophers who made pronouncements by looking skyward were no longer able to advance theories without a solid mathematical foundation.

Is it important for our understanding of the present-day world that an Egyptian, Chinese, Indian, Swede, Christian, Jewish, or Moslem scientist thought and wrote of a discovery first? That the first man to fly was American, a Brazilian, or a seventeenth-century Ottoman? Or that an Indian or Chinese first advanced a particular theory or conducted the first experiment?

It is, and it isn't! Such information is important from the standpoint of knowing the history of how a concept developed. To quote Bruno in my lifetime and to send a "thank-you" to his molecules in the universe might be unimportant to him, but has been very important to me throughout my life. At the very least, he contributed to my gaining power from scientific straightforwardness.

I will shortly test the subject of prejudice. It might be very important for our future for people to understand themselves as a race or herd or as individuals. By overcoming prejudice and looking at the subject free of anxiety, we see that it's not of existential importance to know who discovered what first. But also, it is part of understanding our history and a part of human nature to feel gratitude toward whoever has been helpful to us.

~8~
The Most Gullible Doctor?

Even the most gullible doctor realizes within five to ten years of medical practice that, apart from education and intelligence, there is little biological difference among humans, at least as far as blood groups are concerned. I started to laugh as I typed that last sentence into my computer! I remembered the great Marmara earthquake of 1999, which caused the deaths of more than twenty thousand people. It was the most powerful earthquake of modern times in this area. According to the newspapers, while the Prime Minister was calling for international help, the Minister of Health, a medical doctor, refused the "foreign blood" offered by international aid organizations. He must have thought that "foreign" blood would "harm" Turks. When national leaders have such narrow-minded attitudes, the public is doomed to shed bitter tears for a few more hundred years. Natural disasters like Turkey's earthquake or Hurricane Katrina, which devastated New Orleans in 2005, have been and will be a part of the human story forever. The more we understand biology, the better we can minimize the harm such events are likely to cause.

I'd like to quote a passage from Ludwig von Wittgenstein's *Tractatus Logico Philosophicus*:

This book can be understood only by those who have thought

these thoughts or similar ones . . . what can be said at all can be said clearly, and what we cannot talk about we must pass over in silence . . . I make no claim to innovation, so I don't give it a source, because whether my ideas have been thought by someone else previously doesn't interest me.

As a young man, the clarity of these words fascinated me. But I no longer agree that "what we cannot talk about we must pass over in silence." I believe that by studying molecules, stars, earth samples, and languages, we must investigate the truth and try to understand it. Knowledge, trust, and communication are the basics of our survival kit.

Are people the center of the universe? Today the answer is "certainly not." But the idea that the world was created for man and for his use had been the belief for centuries. How many people do you know who would stand up to declare that a particular hypothesis was wrong and risk their lives for doing so? Aren't we all ready to make certain compromises from what we know to be true when our personal gain is at stake? Because humans are instinctively cautious when they sense danger, they make compromises, and that is why our race has been able to survive and establish civilizations.

The story of an individual who advances truths and is burned at the stake for his beliefs becomes the stuff of legend. Had it not been recorded, thanks to his obstinacy, Bruno would have become yet another of the nameless casualties of history. The ability to compromise might be the reason why we celebrate and admire Galileo, who chose to go along with human nature and stay alive. Many other examples can be given, as the interested reader knows or can find out in a more extensive study of history.

Let's not forget that for many centuries those in power said that "the earth is the center of the universe" and intimidated those who opposed them. It has not been an easy task for humans to overcome this deep-rooted bias.

~9~

Aftermath of the Big Bang:
As the Earth Formed

Astrophysicists tell us the earth was formed about four and one half billion years ago and that a Big Bang occurred around fourteen billion years ago and whose sound can still be heard in space. From the length of the sound waves, it can be estimated how much time has passed. As people continue to observe the universe with increasingly potent optical instruments, knowledge is being acquired at a faster and faster rate.

The year 2005 was a turning point in evolution. On July 4, 2005, the spaceship *Deep Impact* directed a bomb at the meteor Tempel 1. This event is worthy of applause. The idea is to set up a defense against a meteor hitting the earth. Many think it a misguided attempt to keep the nuclear bomb experts employed, even when the nuclear bomb professional seemed less important after the demise of the USSR. When we look at the moon or look at pictures of the earth and see the huge impact craters marring their surfaces, we become aware of how many meteors have collided with planets. Outer space resembles traffic in present-day India, Italy, or Istanbul. Earlier in its formation, meteors repeatedly struck the earth, and scientists believe that basic minerals and molecules carried by these meteors played a major role in the history of evolution. In other words, we are all made of stardust!

Many scientists now accept the theory that dinosaurs became extinct as a result of a meteor hitting the earth sixty-five million years ago. As we saw with reindeer, an unusual temperature change, in this case a harsh winter, and unchecked increase in population can cause death. In 2004, we witnessed the wrath of a tsunami in Thailand and Indonesia, and then we saw Hurricane Katrina. Imagine what a five-to-tenfold more powerful natural disaster could do to the world! People have short memories and tend to forget unpleasant catastrophes, acting as if they never happened.

It is important to get an idea of numerical entities, be it the two hundred thousand years we have been around or the number of people on the earth, or about how long it has been since the formation of the earth.

It's fine to speculate about the universe, but it is difficult to take in the magnitude of the earth's formation 4.5 billion years ago and the approximately seven or eight million years of the evolution of Homo sapiens. The line between science and belief is often blurry and it is easy to confuse which is which. Have you ever counted aloud from one to one billion? Guess how long it would take.

I have made this calculation: If I work at it really hard, I can reach one hundred thousand in a day, a million in ten days. That means, if I don't go crazy first, I will need more than thirty years to reach one billion! Actually, I wonder if it could be educational or even of therapeutic value to count up to a hundred thousand aloud.

This task of counting should be required of every politician who reaches the level of minister or cabinet secretary in any country. This exercise could be done to reach some sort of Zen awareness. It might put an individual at ease and awaken a brand-new respect for mathematics.

This simple counting exercise provides a better understanding of what is meant by the six or seven billion people living on earth. We need to believe the numbers. We could not possibly count the population of the earth ourselves, so here once again is belief in the midst of science.

The calculations required to estimate the age of the universe and the increasing knowledge of outer space have only become possible since the twentieth century and the development of the computer. Mankind is on the threshold of an increasingly rapid age of discovery. It is impossible not to see or feel this acceleration of knowledge.

~10~
The Moon Is as Old as the Earth

At the time of the earth's formation, the moon was two hundred thousand kilometers closer to the earth. Heavier elements, because of their weight, were in the center, and the lighter gases started revolving around the earth's core. The gravity of our planet's star held everything in place, and slowly the atmosphere of gases began to form around the solid earth. At first, the atmosphere consisted of hot poisonous gases. Scientists conjecture that during the earth's first formation, a giant meteor the size of Mars hit the earth, knocking off a large amount of matter from its surface, which then pulled together to form the moon, which traveled in the earth's orbit. The creation of the moon ordered the earth and balanced its orbit. Without the moon, our lava-filled planet, according to the laws of physics, would lurch around the sun, and we would not enjoy regular seasons.

Lava-filled balls evolved into planets as they collided. In the early days, the earth's collision with meteors averaged one a month. How did the seas that cover three-fourths of the earth come into being? It is believed they were formed by icy meteors falling on the earth as well as by steam emerging from the earth's volcanoes, which turned into rain. This happened, according to today's findings, roughly in the earth's first five hundred million years. How did scientists develop this complex theory?

Using the same basic method as is used to measure the substances in human blood, spectrophotometry, astronomers can measure what type of minerals are found in a comet. For example, while most of the water in our seas is H_2O, a trace amount is what is known as heavy water, or D_2O. It is now possible to measure this type of water in meteors as well. Using "tons" of data like this, we have built up our knowledge of our universe over the years.

~11~
Earth Is a Magnet

Lava inside the earth contains iron, which converts the planet into a kind of giant magnet. This magnetic field generally protects the earth from solar winds made mostly of electrons and protons blowing at speeds of 670,000 to 1,570,000 miles per hour. Because the iron inside it has cooled, the planet Mars has lost its magnetic field, and despite obvious indications of having once contained water, Mars now wanders as an extinct planet.

When we say the earth is a big magnet, in a real sense it is possible to find its poles. There is the geographic stationary North Pole and there is also the measurable magnetic North Pole, which is not stationary. In the last century this measurable pole, which had been moving from Canada toward Siberia at an average rate of ten kilometers a year, began moving in the same direction at a rate of forty kilometers per year. Will this trend continue, and what does it mean? So far, no one knows for sure. The balance in which we live is not as stable as many think.

Let us suppose that a period of 4.5 billion years is equal to twenty-four hours, or one day. On this scale, humans evolved only in the last thirty seconds of the day. If we imagine that this thirty-second period ended in the year 2000, we are now in the first second of the new day and will remain in this first second for a few hundred years.

~12~
Sine Qua Non – Oxygen

Oxygen is the sine qua non or "but for which it could not be" of living creatures. I had to learn basic Latin in order to study medicine in Germany. I didn't see the reason to learn a dead language and was prejudiced against it. I was swearing and cursing and very unhappy about my Latin studies. But looking back, these studies enabled me to overcome linguistic barriers scientists like to place between themselves and the general public. And anyway, using the Latin phrase sine qua non makes everything sound more scientific, doesn't it?

For hundreds of millions of years during its creation, through bacterial production, the earth became more hospitable to the development of higher beings. Bacteria started producing oxygen as a byproduct and thereby accounted for the initial rise of oxygen in our atmosphere. Bacteria are fascinating forms of life and can double their population in twenty minutes. They can multiply surprisingly fast, and if conditions are unsuitable, can slow down their metabolism to make spores that can stay alive for hundreds of years. Humans, under the best of conditions, can achieve this "generation time" in twenty to twenty-five years, and that is not a very flexible time frame. While the male can reproduce for a longer period, women appear to reach their childbearing limits around age forty, and the risk of dangerous pregnancies begins at age thirty. There is good reason to believe that this time frame can be lengthened through artificial insemination until seventy or even eighty and beyond. In other words, a bacterium can multiply sixteen times in one hour, whereas one hundred people can become two hundred in twenty to thirty years. If we continue to reproduce and the overpopulation causes an imbalance on the earth, there are many species that can reproduce a lot faster and which could easily take over and dominate the earth.

Bacteria covered the earth, and the oxygen they produced enabled more advanced creatures to evolve. It is unknown when bacteria first appeared on the scene, but this probably can be dated to the first five hundred million or one billion years, evolving either alongside lava, or at the bottom of the sea away from the intrusion of meteors. Because bacteria can be found at the bottom of the sea next to boiling lava, in the darkest canyons, in an atmosphere filled with boiling lava and noxious fumes, without oxygen, we know that they would have been healthily multiplying in the non-oxygenated environment when the earth first formed. Today, the "clean" air we breathe contains between zero and three hundred bacteria per cubic meter. In any fluid, you can easily have millions of bacteria in a cubic centimeter. This is simply another hint about our origins, from the water, if you wish.

Microbiologists can observe the various developmental stages of bacteria just as astronomers can track various stages of the earth's development in the universe. The understanding and definition of microbes in the last two hundred years has clarified our understanding of the world as much as the understanding of gravity has. It helped us get rid of a fair amount of "demons and tall tales." But even in the Middle Ages, people burned the clothes of the diseased and wore protective clothing. They thought "bad air" caused disease. We already had started sensing the world of microbiology.

It's possible to guess a rock's age by measuring its radioactive molecules' half-life, as mentioned previously. Living things entombed during a rock's formation are called fossils. By careful observation, it's possible to find many fossils in stones.

~13~
The First Fossils

The first fossils discovered in rocks are approximately 3.5 billion years old. These are the ancestors of cyanobacteria. These bacteria shamelessly multiplied on the Australian shores, and by multiplying upwards, their remains formed rocks in the middle of the sea.

If we piled up the same number of layers of human skeletons as there are of these bacteria, one on top of the other, the top layer of humans would be living on top of graves 100 kilometers high; in other words, the pile of skeletons would reach into outer space!

Photosynthesis is the process by which plants and certain bacteria manage to survive in the universe by taking advantage of the sun's energy. Chlorophyll is the molecule through which the biochemical transformation called photosynthesis is achieved. Chlorophyll takes advantage of the sun's rays and is the forerunner of the hemoglobin in our blood. The ancestors of these bacteria, through their metabolic waste over a few billion years, raised the oxygen level in the atmosphere from 1 percent to today's 20 percent.

We've already mentioned that any self-respecting bacteria that finds itself in a convenient environment will beget a new generation every twenty minutes. I said earlier that I would mention sex, didn't I? Scientists call this multiplication "reproduction," but in colloquial terms, this is the simplest form of sex.

The molecule described by James Watson and Francis Crick in 1950–53 as the key to life—DNA (deoxyribonucleic acid)—was instrumental in cracking open the secret to life on the earth as well as in further explaining Darwin's theory. The world population was around 2.5 billion when Watson and Crick were working on their contributions. DNA is present in all living cells, and it is made up of four bases (adenine, cytosine, guanine, and thymine) and also of a sugar desoxyribo with a phosphate. DNA exists in plants as well as in lizards, birds, and fish. It is found in every cell; that is, it is found in the cells that make fish swim, in its muscle, and in its brain. When comparing the DNA of ancient inhabitants such as the shark or the turtle with relatively newer species, we can determine when they came to earth as accurately as by looking at a clock (a clock that marks off at least a thousand years with every tick, that is).

DNA contains the key to diseases to which an animal is susceptible and even to what his height will be. The same can also be said for a plant as well as for a flea or an elephant! In addition to the well-known three groups (proteins, carbohydrates, and fats), biochemists view DNA and RNA (ribonucleic acid) as a different fourth group. This main blueprint, the DNA, decides in which sequence the twenty to twenty-two amino acids will be used when building a cell and hence determines just how a human being or an elephant will be formed.

The way DNA is strung together is increasingly understood. For example, the difference between the DNA of a chimp and the DNA of a human is only around 2 percent. Because all mammals share a certain amount of DNA, and primates share even more, the 2 percent that is unique to human beings is enough to differentiate us from chimpanzees.

In 1990, the Human Genome Project was developed jointly in the United States by the Energy Department and the National Institutes of Health. This important project was completed, thanks to the fast pace of technology, in thirteen rather than the estimated fifteen years. This project sought to map and sequence the entire three billion bases of a single copy of the human genome. Now that this project is completed, one can purchase shares of this knowledge bank on Wall Street. This information led the way to a boom in the field of biochemistry.

Allow me to share something about today's knowledge about IQ (intelligence quotient) and DNA: IQ tests are culturally biased. As the mainstream of intellectual capability queried in this test is the "Western" one, Chinese Americans as a group do extraordinarily well. DNA-wise, the

"Chinese gene" is found in Chinese in China, of course, and in many Asian and European countries, but one group with a "Chinese gene" lacking a "Western"-style education does much more poorly in IQ tests: the Native American Indians, though they are geographically located in the western hemisphere. Clearly, a "Western education" is not easy to obtain on the reservations where Native Americans live, not forgetting, of course, the cultural intimidation that has taken place over the centuries.

Let's briefly touch on longevity and genes: Because the fruit fly, or drosophila, has a short life-span and only two chromosomes, it is a useful subject for study in the field of longevity. By methods such as putting the fruit fly on a low-calorie diet, it is possible to extend its lifetime many times over. If the same is true for humans, and if we can find the right combinations of nutrients to ward off disease and promote health, it MIGHT mean that we could extend human life, maybe into the one hundreds with today's knowledge. Would the right diet mean that humans could live several hundred years? From the fossilized bones of early humans, we know that there was no Noah who lived to be 950 years, nor a Methuselah who lived to be 969.

Well, the notion of "year" may have been different then. Such stories are probably the exaggerations humans resort to when they want to make a point. The stories might have been developed with the good intention of recalling the "good old days" and keeping people from fighting and killing each other: in other words, to give them hope.

DNA resides in every bacteria. When I was a medical student, it was believed that E. coli was the most common bacteria residing in human intestines. Later, we learned that anaerobes, or species that survive without oxygen, were in the majority and that E. coli was farther down the scale of intestinal bacteria. This discovery took another ten years to trickle down to most of the medical sector. As a young assistant, I was amazed by colleagues who said, "Those findings you mention are not verified; E. coli is the bacteria most prevalent in our intestines." Well, I wasn't about to start a fistfight with my superiors to defend the honor of anaerobic bacteria!

Everyone has had similar experiences; when you have more insight into a topic that is being discussed and want to share your knowledge, you risk being rebuffed and end up with the appellation "know-it-all." On the other hand, if you have never heard of an expression and miss the point, you get stamped "ignorant" or "dunce-like." Such encounters are how prejudices are formed. We might as well learn to like and cherish the attitude called "prejudice," because we will have to live with it.

One of the reasons E. coli is popular is that it lends itself to being easily produced in the laboratory. Therefore, we know the E. coli species better and have an E. coli-centric (as in egocentric or Eurocentric or America-centric) scientific point of view. Anaerobic bacteria for laboratory experiments require costly technology for the creation of an environment without oxygen.

~14~
Montezuma's Revenge

When one gets the vacationer's diarrhea known as "Montezuma's Revenge," there's no need to fear the always worrisome tropical diseases—often it's just a matter of which subspecies of E. coli will emerge victorious. At one time, scientists believed that examination of E. coli bacteria would determine where feces came from, for example, from which neighborhood in London, as surely as a fingerprint. Later findings showed this belief to be only a cock-and-bull story. Even fingerprints are almost left behind in criminology, because DNA analysis provides much more definitive results. Instead of microbes, we would use DNA now. DNA is a very exciting molecule because it clarifies the lineup of the twenty amino acids that form the foundation of human life.

Amino acids are found in meteors that fall to the earth. Atoms found everywhere in the universe—hydrogen, oxygen (20 percent of the atmosphere), carbon, and nitrogen (80 percent of the atmosphere)—form amino acids under high voltage. Ninety-nine percent of the human body is composed of these four atoms.

When we say "meat," we speak of a food rich in protein. Protein is also found in plants, but to a lesser extent. Amino acids are not living organisms but are indispensable to human life. You get your share of amino acids by simply eating a few almonds each day. It is possible to survive on apples and grapes (with their skin and seeds) without recourse to any other food.

When amino acids form proteins and make up the foundation of a cell, we are talking about living beings. And yet, we've already mentioned that amino acids are found in meteors.

When there is a sudden explosion, as when a meteor hits the world (that is, when pressure is applied), amino acids can form larger molecules known as polypeptides.

We don't know when the first multiplying cell (that is, our ancestor) first appeared on this laboratory known as the earth. But the fact is, multiplying cells came onstage, and in a few billion years, they developed into a more evolved species. When the earth was about a billion years old, other varieties of living things appeared through mutations. And the fish appeared on the scene next.

In 1965, my high school biology teacher told the class that the first fish to come ashore was coelacanth (meaning hollow thorn), near Madagascar.

In 1938, between Madagascar and South Africa, fishermen found a steel-blue, meter-long fish. The head of the Ocean Museum in South Africa, Marjorie Latimer, became interested in this animal, drew its picture, and consulted an authority on fish, Professor J.L.B. Smith, for name and classification.

When Professor Smith first saw the drawing, he said, "I would only have been equally surprised if I had encountered a dinosaur on the street."

Previously known only through its fossils, this fish belonged to a species that lived between 350 million and seventy million years ago, and it had been thought to be extinct. The find generated much excitement in the scientific community, because this species is believed to have been the one that came ashore first. It was later learned that the natives of Madagascar were aware of this "modern science" fish and labeled it as "one whose meat is not very tasty"! Natives often know all about their local environment, but of course, we all know that a species doesn't actually exist anywhere in the world until a "learned" man has seen it and given it a Latin name!

This species was first noted alive at a depth of 198 meters near Madagascar. In 1997, another scientific sensation was the discovery by a pair of British students of a different species at the Manado Tua fish market in Indonesia. This specimen was named Latimeria menadoensis and was also later spotted alive. Nowadays, much mention is made in the press of a fish found in Canada known as the Tiktaalik. This fish, a more advanced phase of the Latimeria menadoensis, can walk on shore and move its head right and left.

Fifty percent of the species that lived on the earth sixty-five million years ago, when a large meteor struck the planet, are now extinct.

If we return to the twenty-four-hour-day example, the ancestors of primates (that is, our species) developed in the last nine minutes, that is, only seven or eight million years ago. And we have been here only some two hundred thousand years (this assumes that Omo I and Omo II found by Leakey in the Omo Valley in Ethiopia or Dali fossils are our earliest yet found ancestors) and are only adhering to the cultural habits developed during the last fifty thousand years, or since the last ice age.

Let's think of population data in terms of length for a few moments:

Think of the time of our evolution of about two hundred thousand years in the metric system and think of one thousand years equaling a kilometer. As the metric system is the predominant scientific measuring system, think of a town about two hundred kilometers (about half the distance between Washington, DC, and New York City, or between London and Paris) away from the room where you are reading this book. Two hundred thousand years ago, when Homo sapiens evolved, corresponds to that distance. Think of yourself walking home from that distance. When you have covered 75 to 85 percent of that distance (that is, fifty thousand years ago, but also as recently as a few thousand years ago) the population was no more than a few centimeters. The year Jesus was born was still hardly at your ankles. When Columbus set sail, you were already home, or only about 600 meters away from the chair where you are sitting now, and the height of the population was barely over your knees. Only in 1900 was the population a bit higher than a well-nourished young man's height today. That would correspond to a distance of about a mere hundred meters away from you after you have walked for weeks or months! Today, that is, only after you have taken a few more steps, the height of the population has reached the height of an average two-story family house!

In other words our ancestors have been on this planet for the last 3.5 billion years and forming cultures on the basis of the distinction self/non-self. Think again of the time you need to count upto one billion! The last 100 years (with airplanes, computers and ipods) correspond to the time we need to count from zero up to one hundred!

It is not possible to speak of the present without mentioning Darwin. And after Darwin, let's look at doctors, lawyers, politicians, and kings in the same breath as microbes, and the all-important artifact—money—so we can look toward a new day.

PART TWO
PECKING ORDERS AND FIRST CULTURES

~15~
Charles Darwin

Anyone showing the slightest interest in science would find it impossible not to be fascinated by Charles Darwin. He not only contributed to biology, but also influenced our understanding of culture and human behavior.

The way any biologist or agricultural engineer understands and interprets evolution today is much more advanced than it was in Darwin's day. He was culturally biased by the concept of races while he developed his theory, but still, evolution as he shared it remains a remarkable milestone in our development. We have to try to understand that he had no concept of DNA, for example, as it was first described one hundred years later. His theory of evolution, however, had a tremendous impact on the way succeeding generations of scientists have viewed concepts of nature.

The majority of scientists rank Darwin's theory of evolution as the most important contribution to the world of science, as it has influenced the way we think more profoundly than anything else. I will try to share Darwin's story, as I believe we are dazzled by many a Nobel Prize-winning novelist, successful businessman, rock star, or President This or Princess That, and Darwin's name is not the household article it should be.

Darwin was born in Queen Victoria's England in 1809, when the world population was about one billion. As a boy, he collected bugs, loved hunting, and neglected his studies. He came from an aristocratic family, and his father, Dr. Robert, was a successful physician who raised his children alone, having lost his wife when Charles was young.

Initially, his father's wish was that Charles would follow in his footsteps, so he sent him to Scotland to study medicine. But Charles could not bear to witness the pain inflicted on patients in the age before anesthesia, and so he cut classes. He was so embarrassed by his failure that he did not go home for a year and bummed around Scotland instead.

When he returned home, his father suggested it might be appropriate for him to become a church minister. Dr. Robert managed to convince his rather lazy, gentleman/sportsman son that the ministry was the best career choice.

"This way," he advised, "you can work on your hobby and continue to collect bugs."

His father went on, "I know you are not dogmatic about religion, most ministers are not, but don't even tell this to your wife. . . . You'll have a respected position in society, a good home."

So his father sent him to Cambridge. At Cambridge, Darwin was influenced by John Stevens Henslow, a minister and botanist. Henslow became Darwin's mentor.

Darwin's religious studies at Cambridge didn't go well. He preferred reading travelogues to satisfy his curiosity and interest in nature. At this time, he fell under the influence of the writings of Alexander von Humboldt, later known as the "father of humanism."

Darwin barely managed to complete his studies at Cambridge and returned home at age twenty-two. While looking for a job, his mentor suggested another occupation. A boat belonging to the British Admiralty, the HMS *Beagle,* was being sent around the world for cartographic studies. Captain Robert FitzRoy was looking for a gentleman who would not only keep him company on this long trip, but record the biology and wildlife they encountered along the way.

At first, Darwin's father was against this idea, but he gave in after Charles's uncle, Joshua Wedgewood, intervened.

Darwin went to meet the man who would become his boss for the next five years, and they shook hands on the deal. Then he received a big shock when he saw the HMS *Beagle.* The ship he was to occupy for the next several years was 27.5 meters by 7.5 meters. He was to spend five years with FitzRoy and sixty others, an astounding number of individuals for the size of the craft, a type of ship with the reputation of being a "floating coffin." Darwin spent the first months at sea suffering from seasickness.

The HMS *Beagle* conducted its first surveys in South America, then sailed on to New Zealand and completed its world tour in five years.

When the ship stopped in South America for its survey work, Darwin took long trips on horseback. He saved the fossils he found and kept a diary describing them. When he encountered the first hunter-gatherer tribes, he noted how these people slept on the ground snuggled up like puppies and how they hunted in herds. He began to develop the idea of a kinship among different races, and this idea continued to develop throughout the trip. On the Galapagos Islands, he noticed how finches' beaks had developed in harmony with the plant life on the islands. He also noticed that animals that had never seen humans did not run away from them.

At the time, correspondence with England took place via ships that made direct trips to and from certain ports. In a letter to a friend, Darwin first mentioned evolution (or in the term of the day, transmutation) and suggested that this idea was as dangerous as confessing to a murder. On one of his port visits, Darwin sent his accumulation of fossils to his mentor Henslow. Henslow not only classified the fossils, but helped spread Darwin's reputation in England.

The Bible claims that humans and animals are God's creation, and the introduction of such a foreign idea as Darwin's could mean being cast off from society. The theory of the day was that God created the world in a week, and man's duty was to rule over the animal kingdom. To believe otherwise was a sin.

On a stop at the city of Salvador, Charles discovered a layer made of seashells on a hill hundreds of meters above the sea. He realized that the earth and the sea were not formed as described by the Bible. In fact, he became convinced that it was formed as described in Charles Lyell's geology book, which he had read.

In 1837, Darwin returned from his trip as a self-assured twenty-seven-year-old. He was no longer interested in settling down as a minister. He gave the scientific community information regarding his discoveries and earned a reputation in London as a noted scientist. For more than twenty years, he kept the idea that was "like confessing to a murder" to himself. He only shared his ideas about evolution with close friends, and he collected data and continued his work. Darwin led this double life for twenty years. He shared his findings without explanation, but fell sick and eventually decided to marry.

Darwin married his cousin Emma Wedgewood, with whom he always got along well. Emma was a religious woman who believed in the hereafter, and theirs was a happy marriage that produced ten children (a rather normal-sized family for that period). Darwin moved to the country and conducted studies on improving dog and pigeon species. He was at that point a well-known man of science. Twenty years went by. He corresponded seriously with other scientists. He led a quiet life as a scientist whose time was consumed with questions of why, where, and how.

In 1858, after hearing that a young man named Alfred Russell Wallace was about to go public with similar theories, he allowed his famous *Origin of Species* to be published. The book gained an instant reputation. Cartoons and articles were published ridiculing his theories, but because of his lengthy and comprehensive studies and the support of fellow scientists, his theory was accepted amazingly fast. His theory, which showed the church's assertions to be at best superficial and at worst fraudulent, was of course alarming. Previously, the French Revolution had upset the social order, caused a change in the ruling class, and generated a chaotic situation.

Following the publication of the *Origin of Species,* Darwin lived for ten more years amid great acclaim. Upon his death, fellow scientists arranged for his burial at Westminster Abbey, where he was laid to rest with great pomp next to Isaac Newton. His biographers noted the

Abbey was honored by Darwin being buried there, rather than the other way around.

Alfred Russell Wallace (1823–1913) is known as the father of biological geography. I'm sure the theory of evolution's relevance would have been recognized sooner or later even without Wallace or Darwin. It is necessary to refer to Henry Walter Bates (1825–1892) in this respect. It was dangerous to put forth such theories, but many were contemplating these ideas. Wallace and Bates took off for the Amazon after being inspired by writings about the formation of the world and its geological evolution.

The Amazon jungle is difficult to convey in words or by viewing on television. It's a corner of paradise where most living things still find shelter. To this day, it is difficult to penetrate more than three or four kilometers a day. Bates remained in South America for eleven years; when he came back, he not only introduced us to more than a thousand new species, but also gifted us with the term "mimicry," referring to a process of evolution that takes place over thousands of years.

For instance, an insect with a red and yellow body tastes bitter to birds, so based on experience, birds do not eat that particular insect. Widely different insects, after thousands of years of evolution, assume the same colors (because birds will avoid them, "thinking" they taste bitter) through continual selection. They end up looking very much alike, although they belong to species as different as horses and cows.

This mimicry was later observed in plants, fish, and snakes. One type of snake adopted the markings of a poisonous snake and is left unharmed by its enemies, even though it is not in the least related to the poisonous snake.

Another example would be orchids, which have parts that resemble food to draw bees. Some fish have undergone evolutionary mimicry not for self-protection, but rather for hunting purposes. A fin in the tail has a wormlike appearance that lures other fish. It is not a matter of intelligence, as found in other insects, such as ants. On the other hand, this can be considered as the "intelligence" nature developed through thousands of years of imitation.

Mimicry found in nature goes hand in hand with the imitative faculty that is the basis for human learning. The expression "monkey see, monkey do" describes this concept well. Communities and tribes learned by observing one another, and the tradition continues to this day in the form of scientific congresses at which knowledge and problem-solving methods are exchanged.

This also calls to mind social mimicry, doesn't it? The behavior of people who create false images of themselves is not very different from mimicry. Think of the huge cosmetic industry. This, however, lies within the scope of culture and not that of evolution. But isn't culture actually a part of evolution? How much of pretentious attitudes are conscious? Deceiving, faking, disinformation, and lying are actually all derivatives of the mimicry seen in nature.

The self/non-self separation is terminology used in immunology. A foreign body is registered by our white blood cells as "non self." The body reacts to this to protect itself. A form of molecular mimicry is when a "non-self body" has some molecules that are much like the organism it "invades." The organism later realizes that this foreign body is alien and must be eliminated, and then develops antibodies against it. As parts of the alien substance are very much like the self, we have molecular mimicry and the possible explanation of autoimmune diseases: a nation's cops fighting its citizens.

As often seen, many people seek a scientific answer to nature's phenomena. Many sense and describe it, and some reach solutions and attempt to share the information, as demonstrated by Bruno, Galileo, Newton, and many more. It is never the work of one, but always of many.

Darwin's theory can be applied everywhere. In politics, daily life, social sciences, and biology, it sheds light on our perceptions. In the medical field, it is the theory of evolution that allows us to explain the reason the AIDS virus or particular germs have gained immunity against certain types of drugs. A virus is neither good nor bad, neither is it an organism that consciously aims to harm humans. It is only reflexively trying to stay alive, and so it undergoes constant mutation to survive within environments changed by medication.

It is wrong to boil down Darwinism to "big fish eats little fish." Viruses are little, but they are able to kill people (who are BIG).

Both humanism and tolerance are reflections of human beings' struggle for survival, as is the wish to kill and annihilate one another by weapons or bombs. We have both the devil and the angel within us. The eugenic ideas supported after Darwin, that is, the idea that humankind can be enhanced just as dogs can, was first proposed by Francis Galton. In *Hereditary Genius* (1869), Galton proposed that a system of arranged marriages could improve the human race. The American Eugenics Society, founded in 1926, supported restriction on immigration from nations with "inferior" stock, such as Italy, Greece, and countries of Eastern Europe. The Nazi practice of exterminating Jews and the attempt to "better" the German race has made this approach a pseudoscience and turned it into a taboo.

The moral taboo of a governmental attempt to organize selective breeding among humans is so abhorrent that "selective breeding" only occurs on the individual level (that is, a bright man seeking an equally bright woman as the mother of his children, and vice versa).

But there again we see that though theoretically we may be able to make taller or fairer or darker humans, the software or the input during pregnancy and the social life of the individual (his software: early childhood experiences, education, the counteraction of traumas) defines who is the smartest.

People sought leadership from the aristocratic class because they expected good governance from those better fed and educated. It was then observed that those who were hungry were more inclined to work.

Those with full bellies showed no readiness to provide for others once their own needs were satisfied. Who are we to count on then? I believe that the present system is good. Because money has no boundaries, when hundreds of thousands of American and European youth decline work they deem below them, those in India and China are happy to undertake the same work. This fact is very well explained in Thomas Friedman's *The World Is Flat*. The financial burden of incarcerating someone can and should become the subject of serious discussion (hopefully, though, never again for political dissidents).

What took place in Hitler's Germany was a misinterpretation of Galton's proposal and an endeavor to annihilate a race. Are Jews a race? Of course not! They thought and perceived that this wide variety of biology was a race, but it was actually a gathering around a belief system. This horrific nationalist holocaust has naturally turned people away from eugenics. A knife is neither good nor bad in itself. It is useful when trimming trees. It can save lives in the hands of a surgeon, but in the hands of a murderer, it is dangerous. This principle holds true for any idea that can be destructive when taken to the extreme. Nothing is good or bad in itself, but in its application by humans and according to humans' intentions, something can be good or evil, useful or harmful. This in itself is the Social Relativity Theory.

In the 1960s and 1970s, doctors saw all germs as enemies and wanted to kill them. Then we understood that if all microbes die, a person cannot survive. That is, the solution often lies in finding a healthy balance.

A person makes a choice when picking a spouse, house, or job. The consultation given parents-to-be about the genetic characteristics of their expected child is a case in point. Today, physicians are able to detect the presence of hereditary diseases in an unborn child during the very early stages of pregnancy. That is eugenics being applied, but in a different sense than was originally conceptualized by those who formulated it. Discarding a notion without considering its positive aspects is a prejudice.

~16~
Monkey See, Monkey Do

History bears witness to inventions resulting from necessity spreading from one culture to another via trade. Sometimes we see an invention in a single culture and sometimes in several simultaneously. For centuries, communities have held secret the knowledge by which they earn their livelihood. This is now futile because of global communication—and so of course, monkey see, monkey do.

For the first time in history, humans are reaching a viewpoint that enables them to understand and recognize racial differences realistically. We're no longer talking of people trying to become acquainted with far-away cultures from the travelogues of a few, but of hundreds of thousands of people of different backgrounds and appearances who, thanks to trade or science, are able to understand the world and its cultures. People of widely different cultures travel around the world, and films shot in any country are simultaneously seen by millions the world over.

People came to realize that their shared values outweigh their differences. They are also beginning to understand that they are on an island with limited means and it is to their benefit to share the accumulated knowledge that will enable them to survive. A medicine discovered in Israel, India, or Sweden is used on a patient in South America in a short time. Scientific knowledge and love increase with sharing. People are starting to realize the need to share this knowledge and compassion in order to survive peacefully and productively on this planet.

It would be wrong to deify people like Bruno, Galileo, Newton, Lorenz, or Darwin. However, as we enter a new age, it is only right to give them their due, just as we credit knowledge acquired from Abraham, Moses, or Jesus.

There are still those who assert that Darwin's theory is unproven. All I can say is, "Mercy!" Look at the theories of the people who make this claim and ask, Which theory enables us to understand natural phenomena and the scientific data we have gathered better than Darwin's? And proof? Well, what more do you need? Maybe a less biased mind and further education?

~17~

First Humans

Myrmelogy is the science that examines tool-using creatures that domesticate other animals and benefit from their yield. The first multistory dwellings are thought to have been built 120 to 140 million years ago. You probably guessed who I'm talking about: Ants!

In order to carry other animals' fluids, honey, or fruit essence, ants take advantage of leaves or sticks (tools) and thus are able to carry ten times more weight. They feed and milk silkworms. We know that octopuses, horses, and dogs, which evolved earlier than humans, have personalities. That is, they have become aggressive or curious or timid by trial and error and hence developed a character of their own. We also know that ants live in a hierarchical society and enter combat. Ants range in length from two millimeters to twenty-five millimeters. We have clues to their making contact among themselves by means of chemicals called "pheromones." Some ants settle down and some migrate. Myrmelogy is a new area of interest. Beginning in the eighteenth century, it has rapidly become a new science.

Another species evolved on earth more than a hundred million years after ants. Homo erectus is the first species that can live erect on two feet. Running on four feet has its advantages. One can continue to escape if one foot is injured, for instance. The advantage of standing on two feet is that the hands are free to use as tools and to carry things. Man started making tools out of stone only four hundred to five hundred thousand years ago.

The remnants of Homo sapiens neanderthalensis, an important representative or subspecies of the human species, are found all over the world. Homo sapien sapiens, that is, modern man, is also found all over the world. Our species is only about two hundred thousand years old and is known to have existed for over a hundred thousand years simultaneously as the Neanderthal man.

We know that Homo erectus migrated from Africa to Eurasia a million years ago. There is evidence that five hundred thousand years ago they spread from Java to Europe and, besides knowing how to carve tools, were capable of laying stone pavements between two adobes and lighting fire. As the male was, and is, usually physically stronger, it was the male-dominant era of human development. Even today, males will want to control the fire at grill parties. You might be hard pressed to find a woman at the grill!

History is witness to the two types mentioned above: our ancestor, the modern man whom we know from drawings on the walls of the Cro-Magnon caves, and the Neanderthal man, who is a little shorter but stronger. Cro-Magnon takes its name from the cave in France where the drawings are found, and Neanderthal from the town in Germany where the first bones belonging to the Neanderthal man were excavated.

These two evolutionary relatives, like horses and donkeys, are from the same part of the world. The last remnants of Neanderthal man found so far date back thirty-two thousand years. We do not, of course, have any evidence of how their species became extinct. It is not biologically likely that they became psychologically depressed and decided not to carry forward their species. What is certain is that they entered into combat with other species and that our ancestors won out. We do not know whether modern man ate the Neanderthal man or not. Looking back in history, it seems normal that they did so. (Nothing is alien to humans when physical or pathologic psychological necessity dictates it.) It is likely Neanderthals became extinct because of harsh weather conditions, at which time their hunting skills were insufficient for survival, and that somehow our ancestors brought about their end. Neanderthals and Cro-Magnons could possibly even have mated and had offspring. Considering man's impulse to mate with every creature that comes his way, this is highly probable. Think of mules, which are small when the mother is a donkey and big when the mother is a horse. Mules cannot give birth, so a mule is a biological dead-end.

What we can detect of the story of modern man begins about fifty to sixty thousand years ago. The modern man who advanced as far as Australia makes drawings as fine as any artist today. The tools he makes are finer and more usable compared to those made by First Stone Age people. Today, we only have access to clues found in caves, because only they were preserved. It is not hard to guess that they usually lived out in the open. Because of natural decay, there are few traces left of what they produced out of wood or leather. The only thing we know is that they brought the raw material to make paint from far away, if need be. Our sense for aesthetics must have developed early. The blue mosaic floor tiles (lapis lazuli) found in Roman houses in Anatolia came from what is now Afghanistan. Long before that, lapis lazuli was used in ancient Egypt as an ornamental stone. Stone Age people fetched the colors used for paintings from hundreds of miles away. Don't we move tons of earth to get to gold nuggets? We know, for example, that humans made flutes out of swan bone and elephant tusk about thirty thousand years ago and shoes some forty thousand years ago. At least that is what the findings suggest. It is not hard to imagine that people made shoes a lot earlier and that we will find some evidence for this in the future.

People living today are of the same species. People living on different continents can have children with one another. All of us have evolved from the same ancestors. Migration holds an important place in the evolution of mankind, as it has meant learning different approaches from other cultures. Another important milestone is the agricultural settlement of hunter-gatherers.

It is widely accepted that the supremacy of the Cro-Magnon man over the Neanderthal rests in the Cro-Magnon man's being able to talk and communicate better. Except for the Neanderthals, there were some other hominids, like the Georgia man or Handy man (Homo habilis) or Flores man, which have all become extinct. The findings about these species are either rare, as in the case of Flores man, which is believed to have lived some eighteen thousand years ago, or older, as in Homo habilis, which is dated back some two million years.

Anthropologists say that the ability to produce more words was the result of a mutation in the larynx and that this brought with it a far better quality of communal work, communal hunting, cooperation, and settlement. It was this supremacy that most probably brought about the end of the Neanderthal man.

Africa's San people, the hunter-gatherers once known as the Bushmen (which some may remember from the movie *The Gods Must Be Crazy*) carry more ancient evolutionary lineages in their DNA than any other people and exhibit a direct living link to Adam. Their language, Xu!, is another link to our ancestors: It is characterized by the clicking sound and has over one hundred distinct sounds, whereas most of the world's languages, including English, use only twenty to forty sounds.

The ability to reach a detailed understanding through different sounds brings an important advantage. This advantage is still used today in a more refined way, be it between cultures or between castes. Scientists generally prefer to use terminology that ordinary people cannot understand. This is partly necessary so they can share knowledge, including nuances and details. Partly, however, it is merely used by members of a particular occupation to establish the security of their position within the masses.

The first Stone Age people knew how to count and they had a notion of numbers. Numbers as tallies have been found in caves, dating back fifty thousand years. The first notion of written mathematics is attributed to the Sumerians eight thousand years ago. You would not worry about counting if you didn't have possessions or want to construct something.

Once something is grasped by trial and error, then it can also be taught. Animals have numerical abilities—not only primates, but also raccoons, rats, and even birds. They can do simple addition and subtraction. We know it through a variety of experimental settings, and this is one of the reasons why I keep saying that science is fun. Think of yourself as a Stone Age person for a few moments: Your prime interest would be eating, then shelter and sex. You would soon find out that the sea is the prime source of food and learn to make some sort of wooden boats. You would start using numbers, unconsciously counting your fingers and the animals you catch, and you would start communicating the numbers soon after.

~18~
Monkey Education

Monkeys' voices were recorded in fieldwork carried out in their natural living environment in Africa. Their effects were evaluated by listening to them over and over again. There were sounds that warned against dangerous animals, as well as sounds that indicated "big bird but not dangerous," meaning "no worries." The sound produced while approaching an inferior animal was different from that produced while approaching a superior one. The sound that warned against a snake was different and warned the monkeys to run toward open space, whereas warnings against birds told them to run into the shrubs. It was understood that baby monkeys learned this language by the age of two—that is, they could not produce the sounds properly or understand them at first, just as with children during the first years of their lives.

Language is used today not only for communication, but also for exclusion. For instance, scientific language is used to emphasize and preserve the autonomy of secondary branches, or "new" sciences. It is also used to exclude the public. For example, an infection caught at the hospital was called "hospitalismus," that is, an infection that the patient caught at the hospital. Later, when more people understood what it meant, it was called "nosocomial infection."

The human animal needs an education period of about twenty years, or sometimes even more, to become functional in today's society—a serious bit longer than a monkey. But the similarities aren't only in the learning of language.

M. Keith Chen of Yale-New Haven Hospital conducted studies with capuchin monkeys. When taught to use money, a group of capuchin monkeys responded quite rationally to simple incentives, responded irrationally to risky gambles, failed to save, stole when they could, and used money for food and, on occasion, sex. In other words, they behaved a good bit like the creatures that most of Chen's more traditional colleagues study: Homo sapiens.

The introduction of money was tricky enough; it wouldn't reflect well on anyone involved if the money turned the lab into a brothel. To this end, Chen has taken steps to ensure that future monkey sex at Yale occurs as nature intended it.

I believe that after speech, the most important development is the invention of writing, which dates back only six thousand years. For hundreds of years, as writing evolved, it was mainly used for keeping track of rulers' possessions and then to write down rules and laws. Writing basically spread by the monkey see, monkey do principle, modified by many cultures to underline the uniqueness of one culture from others by its alphabet. Unique in the real sense of the word (that is, without copying from another), alphabets started evolving in the Americas thousands of years later.

Fifty thousand years ago, during the ice age, when the lands between Java and Indonesia were nearer each other, people sailed to Australia, and at least about seventeen to eleven thousand years ago, they went to the Americas, crossing the Bering Strait, which was then a land bridge covered by ice. It is not known whether the first people who settled there stayed on the west side of the Rocky Mountains and the Andes, that is, whether they spread from north to south only or to the east as well. New information that comes our way each year enables us to understand these migrations better.

We have a relatively long silent portion of human history until we were able to write. For this period, scientists have to make guesses based on a bone found here or another there. What is known is that for fifty thousand years after their migration to Australia, these people lived separate from the people who lived in Europe and Asia.

The people who went to the Americas were cut off from others until Christopher Columbus arrived. This, at least, was our understanding until a theory by a British retired submarine commander, Gavin Menzies, launched the theory that the Chinese beat Columbus by about seventy-one years.

He believes that the world and all its continents were discovered by a Chinese admiral named Zheng He, whose fleets sailed the oceans in 1405—a notion that is partly documented in Chinese historical records. The existence of maps in Europe that showed Africa and the Americas before Columbus and other European seamen had "discovered" those places suggests that someone was there before Columbus. Menzies believes that those were copies of Chinese maps. In 1424, when a new emperor decided to turn inwards, old records were burned and China wasn't able to follow up on this "discovery." The Ottoman admiral Piri Reis's map of 1513 shows America, thereby challenging Columbus's discovery of America as well. In his notes, he states that he has copied his map partly from maps that were over a few hundred years old. Until new scientific evidence is found, this map quest will keep fascinating researchers. The anthropologist Thor Heyerdahl and his crew proved with their "prehistoric boat" *Kon Tiki* in 1947 that it was possible to reach Polynesia from South America with primitive vessels. Though new evidence shows that Polynesia was settled from Asia, they made the point that prehistoric sailing over thousands of kilometers was possible. In 1970, he was able to cross the Atlantic with a papyrus boat, *Ra*.

~19~
First Steps toward Religion

Excavations show that Stone Age people buried their dead. This was first done by the Neanderthals to ennoble the dead. According to scientists, the habit of burying the dead brought about an increase in the development of religious sects as well as belief in an afterlife. There must have been a well-defined pecking order between the people from the early ages on. Some were buried with many flowers and precious stones, and later with much gold, and some were buried without any ceremony.

Today there are people who still live as hunter-gatherers. Some tribes migrate seasonally, spending the summer in the mountains and winters in the valley. The numbers of these Stone Age-like people are decreasing rapidly, mostly because of contact with civilization. We know that these people know their surroundings well and have a better notion of plants and animals than many "successful" city dwellers. One of the earliest monuments Stone Age people built is in Göbeklitepe, dating back some 11,500 years. It is near the Iraqi border in southeastern Turkey. Archaeologists first thought the monuments to be from the Roman Period (about ten thousand years later!), so fine are the stone-carved figures. With modern techniques, they were able to date the worship grounds of these people more exactly. There is no pottery; pottery comes later. They probably used some wooden vessels, but nothing has been found because of natural decay. We can easily assume that our ancestors first carved wood, then stones. After a while, they started forming clay and heating it, making vessels for fluids and food. The transportation of heated clay was the first step toward pottery, probably nine or ten thousand years before the present. And new finds are pushing these dates back further and further.

Let's go back to the origins of human cultures: When such fertile valleys and woods exist in the Americas, why is it that the most important cultural developments took place in Mesopotamia and the Near East? How is it that the Conquistadores overran America with a handful of people with gunpowder, swords, and horses? Why had the cultures in the Americas remained backward, without gunpowder or the wheel?

The first farming communities developed in places where the geography was conducive to farming, where grain varieties were most plentiful and biology was at its richest. Animals and plants were domesticated in the geographical region where man could find the greatest variety, that is, where there were alternatives (that is, in the richest testing grounds). Mesopotamia is hence the motherland of both wheat and barley, still the most consumed foods on earth.

Human domestication of plants can be seen in three stages: gathering, in which people gathered annual plants from wild stands; cultivation, in which wild plant genotypes were systematically sown in fields of choice; and domestication, in which mutant plants with desirable characteristics were raised. Cultivation is the essential stage, as the repetitive cycle of sowing and collecting of wild plants gives rise to genotype accumulation that leads to domestication.

New archaeobotanic evidence seems to indicate that fig cultivation was widely practiced in the Near East twelve thousand years ago, nearly a thousand years before the domestication of cereals and legumes. The domestication of the fig preceded that of other fruit trees, such as the grape, olive, and date, by almost five thousand years. In North America, indigenous plants like squash and sunflowers were domesticated five thousand and four thousand years before the present.

We believe that there are between 3.5 and twenty million species on earth today. If we can determine the age of a piece of wood or stone or fossilized bone accurately, why do we speak of such a wide range of biological species when it comes to living creatures?

The answer is simple. Man is an interest-guided animal who seeks easy solutions. Our narcissism knows no bounds; we are not as intelligent as we like to believe. We do not want to work or think if we don't have a serious problem. We prefer to use as little energy as possible.

We prefer enjoyable matters. We have lived as tribes for security reasons for millions of years. The reflex to live together is still the main reflex that determines our choices and behavior. If you are part of the herd, you are "self"; on the other hand, if you are not part of the herd, you are "non-self," and there is danger in that.

If a human can only define himself as part of a herd, he will be ready to die if his herd, be it of cultural or national definition, is threatened. And the biggest "herd" is probably a religious one. More wars seem to have been fought in the name of religion than for anything else! We must encourage a person's development of a sense of self and a hope for the future before he or she will be able to think of humanity as a whole.

While living as herds, the principle of the pecking order, as suggested by Konrad Lorenz and other scientists, is effective. Alpha animals fight to the death in order to attain power. As long as they have power, they can mate with whomever they want within the herd. This is an example of how the genes of the strongest become widespread.

In the pecking order, when the alpha animal is fed and no one is threatening his authority, he is a prime example of untroubled tolerance. Young and sick animals' untoward behavior does not bother him. Animals that can never become an alpha animal are also tolerated by the leader. Isn't there a parallel between the tolerance shown to the court jester and that toward the absent-minded scientist? But when the alpha animal feels his position threatened, he is merciless. One of the main principles in training a dog is that if your dog acknowledges you deep down as the alpha animal, he will not object even if you remove his food while he is eating; but if there is not full acceptance, your act might be met with growling and biting.

I believe that animals such as ducks, dogs, or horses were domesticated in ancient times by taking their babies from their mothers and having them recognize a human as the alpha animal. To our forefathers, for a very long time, it would have been easier to simply kill and eat these animals. Then domestication "evolved" with need. As the status quo with the domesticated animals was accepted and widely used, it led to new developments, such as, for example, the invention of the wheel. If no animals were there to pull a cart, there would be no need to invent a wheel. In some continents, the wheel couldn't be invented, because when humans reached that stage, there were no animals to domesticate. Camels and horses, which also evolved in the Americas, were extinguished long before humans could domesticate them.

Pecking order is an important biological principle. It applies to human herds and our hierarchies in this day and age as well. As herds expand, it's becoming increasingly difficult to guess where one stands in today's cultural polyphony. Should one be impressed by someone's cell phone, car make, or hairdo? Other important criteria are what country's passport one carries, as well as a person's age, height, and weight.

Certainly the pecking order that is easily applied in a tribe of forty to sixty people becomes more difficult as the population increases.

Next to IQ, which stands for intelligence quotient, there is EQ, or emotional quotient. I hope that soon a BQ, or bias quotient, will also be measured. As we will talk about later, bias is part of any culture, and therefore is important to identify, define, and measure objectively. At some point, we've got to reconcile with the pecking order. It is mostly by chance that we live, are educated, and choose our herds, our friends (birds of a feather flock together), and if we have the means, our country of residence.

Every person, upon meeting another, runs him through the useful/useless or self/non-self test, or in computer language, a "1" or a "0." We do this unconsciously; those who deny this are only fooling themselves. Each one of us perfects this basic approach according to the level of his or her culture and education. Everyone is biased in a personal way. When deciding whether a meeting will be of benefit from a sexual standpoint, idea standpoint, or the standpoint of having a good time, the decision is reached semi-consciously and the prejudices an individual has acquired through family and culture will affect this decision. These will increasingly acquire mathematic definitions. In the future, bias will be measured and eventually an equation will be formed that indicates "x" (bias) goes toward infinity but cannot reach "y" (zero or no bias at all). Humankind will only approach obvious facts and will only be able to hold onto those facts through intuition.

I mentioned approximately ten million biological species (a range of 3.5 to twenty million) and that we could not give a more definite figure. This dubious estimate of biological species shows our ignorance on the subject of general biology, and also that science is not as "scientific" as is generally thought. This number also presents a challenge for younger generations and shows that serious, well-established elderly men and women of science aren't as full of wisdom as they might think.

Let's briefly look at popular books that have shaped our understanding of the human animal:

Zoologist Desmond Morris's book *The Naked Ape* popularized anthropology and human evolution. Leon Festinger's book *The Human Legacy* is another book on this subject. The writer, whose background is in psychology, researched man's evolution and the cultures of Eurasia.

Another classic is *Selfish Gene* by Richard Dawkins, published in 1976. Interestingly enough, Dawkins says that "nice guys finish first." I also believe that people who are "good" (that is, able, talented, strong, or intelligent people who have realized their interdependence) will be more successful.

Ornithologist and UCLA physiology professor Jarred Diamond has written bestsellers about human beings and our culture. In his book *Guns, Germs, and Steel*, he describes the "evolution of government and religion" and calls the advancing new classes "kleptocracy."

Cyril Northcote Parkinson's (1909–1993) ideas about bureaucrats are becoming popular: If they "steal" from their "work time" or during their distribution of knowledge or goods, or if they lobby for new subordinates and thus indirectly secure their jobs, they end up receiving more than what society is willing to pay for the job. This "white collar theft" is getting more and more attention from "normal citizens."

In the final analysis, life is a race. It starts with millions of spermatocytes racing each other to survive; only one or two succeeding if the receiving side is "ready." All others die after each ejaculation. We later race to get food, get the "best scene," or help the "best leader" to come to power, to get the best house, to give our children the "best education" or our parents or grandparents a place in the retirement home we think is the best. So it is only natural that we like watching "contests" (that is, other people racing), as this is the central biological theme in our lives.

Thinking along the same lines, we also should consider addiction in history. As someone who was addicted to tobacco for a long period, and who loves to play bridge, I think this is a central theme. Someone who consumes tobacco is actually sending two basic messages: The first one is challenging death, though we should not forget the momentary pleasure one gets out of it, and the second is a suicidal motivation. Next to the physical addiction, there are surely mental components in addiction.

If addictions and games are one major theme of human behavior, another is prejudices, or generalizations (laziness of mind or getting stuck in the status quo). All make life easier and are addictive human behavior.

~20~
Games, Depression, and Targets

Games, addictions, and prejudices simplify everything. They entertain the mind, which basically is very lazy. We have to accept this laziness as a part of our biological system. In my opinion, human beings function using approximately 20 percent of their intelligence. Only when an urgent situation occurs do we tend to awaken from our slumber. Even then, man is not a very intelligent species when one looks back at the pace of development.

I believe that at most, 30 percent of human beings actually work. Many are retired (and this portion will become larger and larger), some are in the educational process, and some are sick or unable to work. May I be allowed to be biased enough to believe that a majority of bureaucrats, be they civilians or military or clergymen, seldom work? Imagine: These people get paid month after month no matter what they do. Even the best would deform and start thinking about fringe benefits and optimizations under such circumstances.

Despite all the problems that I have been talking about, people who work creatively are the reason why I am able to look at our future with hope. And I think they are numerous enough to achieve a happy ending for humanity. Laziness, having fun, sleep, and depression are all part of human biology. When we think of our close relationship to animals that spend the whole winter hibernating, I think laziness is a healthy part of our biological heritage. Every human being gets depressed at some period in his or her life. Antidepressant medication is one of the most used of all medications. Not only aimlessness, but also bleak prospects in the job market or troubles in relationships can be triggers for depression.

When a sensitive person gives some thought to things happening around him or her, it is not surprising that he or she becomes depressed.

In order to get out of this depression, one needs to find a goal. But human beings usually prefer to live their lives going downstream. Only a few try to swim against the tide, as this would mean reaching out for others, with all the inherent dangers of doing so. Societies traditionally prefer the easy way out. In democracies especially, this could even become a systematic bias. Humans have tried to "own" things that by definition cannot be theirs. We have even tried to create laws, or religions, to justify our behavior. We have consumed a significant portion of our habitat, as has our relative the grasshopper. On the other hand, man is not foolish enough not to sense his unfortunate end. What can he do other than have fun, spend his time playing, or become depressed?

When we look at the careers of successful people, in sports, business, or science, we often witness an ever-growing greed. The same urge can explain the success or the failure of certain tribes or even empires. There are always balancing natural factors such as competition or oversaturation that cause a halt or regression. It is different with our war against nature. We have only become aware of the destruction we are causing in the last decades. The amount of concern for the environment has been increasing at a pace that is encouraging, yes, even fascinating for me. Capable people live longer, and once the worldly concerns have been satiated, people will start to set new patterns for their surroundings. Once patterns are set, masses will gravitate to the cause.

In the mild climate of the Mediterranean, where winters last only a few months, people have been deforesting their habitats without giving it a second thought. Mainly because of longer winters and the need for heating, it was the people who migrated to northern Europe who started the tradition of managing and replanting forests.

We prefer to live and play without thinking whenever we can. We can also observe this never-ending playfulness in animals. When their stomachs are full, they like to play with each other. Every dog owner can see how puppies play with each other and with their parents. War games are the best training in learning how to survive. We can also observe this in cats, horses, and hamsters. The play instinct of the fed animal can be translated into gambling saloons and Internet games for humans. Racing or having a representative race for oneself is popular. About 30 percent of the world's population watches the finals of the World Cup soccer championship.

When a goal is set, people will run toward it thanks to the herd instinct. One needs to believe in his or her goal. Many examples have been given in this respect. Religions are the best proof of it, the goal being the kingdom of the superior race in one or a place in heaven in another.

Communism and Nazism were other examples of totalitarian religions. Conquerors such as Alexander the Great, Attila the Hun, Genghis Khan (whose kingdom was twice the size of the well-known Roman Empire), Caesar, and Napoleon are proof of what human beings are capable of achieving. If we don't want to become extinct, like many species on this island planet, it is the responsibility of the fortunate who can understand this to agree on a lowest common denominator to end our wars among each other. But more importantly, we must agree on a way to end the idiotic war we are fighting against nature.

~21~
Dogs and Horses

It is assumed that the first dog was domesticated about one hundred thousand years ago. The DNA analysis of American dogs gives definite evidence of Asian ancestry. The different pedigrees we know today have been bred within the last five hundred years.

The domestication of pigs and goats has taken place in the last eight to ten thousand years. Presumably, chickens were domesticated around the same period in Asia, later to be adopted in Western Asia and Europe.

After the ice ages, we see that people go into a more sedentary lifestyle, and we start seeing the first domesticated animals. We call the period between 7,500 and 4,300 years from our era, in other words, up to 2300 BCE, as Neolithic. Of course, these dates are subject to change as new evidence arrives from related fields. And maybe in the near future we will find ourselves classifying these periods as those of the dog, mammoth, sheep, pig, or horse.

We see that there is extensive trade and production of pottery in the Neolithic Period. Walled cities are being built. The period from 4,300 to 2,800 years from the present, that is, 2300 to 800 BCE, is called the Bronze Age. Trade became even more important. Besides cups made of wood and terracotta, we find the use of bronze in protective gear. Fortification sites became common both in the Near East and Europe. The great ancient Egyptian monuments (like Luxor and Karnak, the greatest temples humans have ever erected to worship their gods) were built in this era. It is fascinating to see the pyramids and such temples in the middle of the desert. If you imagine how many millions of tons of wood was needed to erect these monuments, and bear in mind that planned "re"forestation is a human achievement dating back only a few centuries (planting orchards dating back a few thousand years), then the reason why these great monuments lie in the desert hits you. People cut down all the trees near these temples, so it became unfeasible to live in those surroundings and people left the monuments to sandstorms. The land was overused. They were not able to pay bureaucrats, and herds big and small left for more fertile regions with new trees to cut and new monuments to be built. The story of ancient Egypt has been replayed within my lifetime in Ethiopia: In the fifties and sixties, "The Empire's" forests decreased, in the last decade alone from 14 percent to around 4 percent. The region has been the theater for bloody wars, massacres, and droughts killing hundreds of thousands, and has become one of the poorest and most problematic regions of the world today.

The Bronze Age is also the time when people called their leaders gods, and these leaders, after erecting temples, called themselves gods (Ramses II, for example). Egyptian gods were smaller than the emperors in stone carvings. Then Pharoah Akhenaten started worshiping only one god, "Aten" (or solar disc). A picture of him kneeling before Aten can be found in the Egyptian Museum in Cairo. This one-god notion did not last long in Old Egypt, but it influenced much of the history until present. The legendary figure Moses seems to be associated with Akhenaten. After Akhenaten's death, a famous follower was Tutankhamen, previously named Tuthankaten, but then renamed after the Amen priests, the carriers of the bureaucracy and state who preferred the multi-god system. Tutankhamen became famous posthumously, a child Pharoah whose gold- and gem-rich burial site was the only one discovered intact in the Valley of Kings (the discovery was made in 1922 by the British Egyptologist Howard Carter). Except for the fights around the creation of the single-god entity, the Egyptian men of cloth (that is, the religious bureaucracy) were probably the main reason why tens of dynasties were able to rule in ancient Egypt when compared with the world powers built by Alexander the Great or Atilla the Hun or Genghis Khan, which all blossomed and collapsed more or less within a lifetime of one warrior king.

The Iron Age, the era of warriors, continues from 800 until 15 BCE. Trade was very common; one encounters traces of fire and destruction in many settlements. This was possibly a period when many wars took place between tribes. The period between 15 BCE until the fourth century CE is known as the Roman Era. This period is followed by two centuries of constant migrations. At this point, one needs to mention the person who built the greatest European empire in history, Attila the Hun. Within a period of twenty years, his rule extended from the shores of the Baltic to France and the Black Sea, and ended with his death in 453 CE. Although he has been mentioned as a great leader by some, his name rings bells in many minds as the "barbaric conqueror."

I would like to recommend that the reader consult encyclopedias or the Internet about Old Egypt, the Hittites, Greeks, Lydian, Chinese, Huns, Mayas, and Incas. Read about the geography and cultural history of whatever civilization comes to mind. The terms "Bronze Age" and "Iron Age" are actually a bit outdated, as the Americas did not reach these ages, but the Incas, Mayas, and Aztecs were otherwise widely developed.

The period between 600 and 1500 CE is known as the Medieval Period. The turning point here is without doubt the discovery of America, the catalyst being the fall of Constantinople in 1452. This period also gave birth to the Renaissance, religious wars, and the Reformation. After this, modern times arrived, as did Eurocentric dominance of the Christian culture. Just as history is written by the victors, science is also directed by the "superior ones." How or when future historians will end this period, we don't know yet. It could be the Second World War or the fall of the Soviet Empire or the birth of Internet technology.

In the last century, science and anthropology have become increasingly America-centric. It is fortunate that this center has shifted from Europe to America, with a more neutral approach to known and well-recorded history. Results of American archaeological, anthropological, and medical research, as well as space technology, dominated the human legacy. Western history books still mention the Chinese, Japanese, or Ottomans with only a few (mostly biased) sentences. Just look at any high school history textbook. Though Chinese culture is one of the richest originators of human civilization, it has been the privilege of the learned few to know. Today, however, these balances are changing. The reason is not a cultural one, nor that of a superior system. The reason for the recent success of India and China is a biological one, and a simple one at that. Hungry people with not much choice tend to work harder than the well fed. Well-fed people and castes with routines stabilize the herd and keep the status quo, but don't like working and risking much for new ventures.

Did the Europeans discover the Americas, and not the Chinese? We don't know for sure, as mentioned earlier. If it was China's privilege to discover the Americas before the Europeans, it occurred within an order-command sequence and was directed from a central power, with no capability of adapting and surviving in the New World. Some sailors probably reached America before Columbus, but like the number of bacteria needed to cause Montezuma's revenge, their number was not large enough to change things and keep the connection to the Old World. The Europeans who settled in the Americas and later "imported" slaves from Africa had adventurous souls and left a chaotic continent with wars and famine behind. They had no other choice than to survive on the new continent. Why didn't the Native Americans or Mayas or the Incas "discover" Europe? The answer to this question is simple: They did not have the biodiversity of Mesopotamia or the constant religious wars that forced them to further development or the fruitful trade of ideas and science that took place over the centuries on the Eurasian continent.

~22~
Why Domesticated Animals?

The role of domesticated animals in the history of human evolution is immense. The reason why humans started to settle and farm in the area between Euphrates and Tigris is because of the biodiversity there.

Horse, cattle, goats, and sheep were only to be found on the Eurasian Continent when human populations began settling and farming.

The domestication of the horse in human history is comparable to the change of Cinderella's pumpkin into a horse-drawn carriage, like a stimulant playing the role of booster in a chemical reaction. An informative source about the legacy of the horse is the International Museum of Horses in Kentucky (http://www.imh.org/imh/imhmain.html). There is evidence of the domestication of the horse as early as six thousand years ago. Before that, we know that horses were hunted.

The invention of the wheel falls into the same period as the use of oxen (castrated bulls) and horses. The impact of the domesticated horse has been immense, and each technological step in their domestication, like the horseshoe and stirrup, has probably had as much of an impact on humans as the development of tanks and airplanes. We believe that these achievements came from Asia, with the Turkic tribes and Mongols or Skythes. The horseshoe and especially the stirrup made a war engine out of the horse. The rider was able to stand up while riding and use all his weight to throw a spear or use his sword, and I believe the widespread use of this made all the difference for the invasions of the Huns, Mongols, and Turkic tribes until the first millennium of present time.

We realize why Native Americans did not invent the wheel. There had been no need for a wheel on this continent, because they did not have the horse or ox to pull it. And the early Americans were fortunate to live in a much richer continent, with a much smaller population. In South America, the native leaders had only their slaves to carry them on sedan chairs. It is not hard to imagine that even slaves were not docile enough to pull carts like oxen. Native American societies did not work with bronze or iron, probably because they simply never reached the population densities reached in the Middle East.

Were there horses in America before the European settlers? As we know, the horse was introduced by the Conquistadores after the discovery of the Americas. This is when the Native Americans first saw a horse. Some of these horses fled and became wild, even forming wild herds. These are known as the Mustangs. Having tamed them, the Indians later started using them.

There have been horse fossils encountered in the Americas as early as nine thousand years ago. We do not know how these horses became extinct. As they were hunted for their flesh in Europe, they probably were also hunted to extinction on this continent. But it is certain that when Native Americans reached the level of being able to domesticate the horse, there were no horses left on this continent.

The following is necessary for the evolution of cultures:

1. A certain population density so humans can interact.
2. Suitable plants for farming and farmland.
3. Suitable animals for domestication.
4. Interaction among cultures to speed development.

Let's get back to the evolution that took place in the Old World:

Distances were shortened through the use of horses. Trade, transportation, and agriculture was revolutionized. For the hunter-gatherer to settle, the sedentary lifestyle needed to offer advantages. The first advantage was security. The second was the surplus of crops when men plowed the fields with the help of animals.

The number of animals that human beings can tame and which offer meat, milk, and skin is limited. The animals suitable for domestication are but a few: sheep, goats, dogs, horses, pigs, cows, donkeys, and rabbits, and among birds, chickens, geese, turkeys, and ducks. Alpaca and buffalo are not completely suitable for domestication. Elephants, which in the past have even been used in battles, are still to some extent used in Asia, although they can't exactly be called user-friendly pets.

Other than the yak or reindeer, animals that can be used only in certain climate zones, only the previously named animals are suitable for domestication out of the 5,500 mammals left on our planet, and only those few previously mentioned birds out of ten thousand bird species.

More animals are not fit for domestication even with modern methods, as their genetic heritage won't allow it. On the American continent, besides the dog, only the guinea pig and llama and its close relative the alpaca are suitable endemic animals for domestication. None of them is suitable for transportation or plowing. In Australia, there are no native animals that can be domesticated. Fancy trying to plough a field with a kangaroo! Even the modern Aussies would need a few more decades of evolution to develop a pull cart for the kangaroo, along with suitable roads. In Africa, we need to differentiate between the South ("Black Africa") and North ("White Africa"). The Sahel Zone divides the continent in two. Sahel is a desert zone with an area of five thousand by three hundred kilometers. Even with modern motor vehicles, it is a difficult and dangerous undertaking to cross. Until recently, for practical reasons, North and South Africa were separated. In the equatorial region, human beings have developed their dark skin as protection from the sun. The same thing can be observed in southern India. The gene commonly found among southern African populations, known as sickle-cell anemia, is also the natural protection against malaria. We still do not know in detail when and how the evolution of the different human races has taken place.

Two totally different cultures evolved in the southern and northern parts of Africa. North Africa, as a part of Euro-Asia, is home to the highest number of animal and plant species and is the largest land mass on our planet. The highest number of animal species has evolved on this continent, and despite human populations, has continued to survive. The South of Africa, on the contrary, was a small, naturally encapsulated world of its own until the last few hundred years.

Paleobotanists (the scientists who study the past of plants) have found, similar to domesticated animals, that most of the crops used today originated in Mesopotamia.

Plants originating from the Americas that are suitable to farm and are sources of carbohydrates are maize (corn), potatoes, yams, and zucchini. Tomatoes, a native of the Americas, are not a significant source of carbohydrates. Tobacco, cannabis, which is the source for marijuana, and coca are also from America and have no nutritional value. Potatoes cultivated in what is today Peru/Bolivia had only spread in South America along the Andes before the Conquistadores' arrival, who with their horses and ability to cross natural borders took this plant to North America and to Europe.

Looking at the geography of the Americas, we notice the axis of this continent is in a north-south direction. Different cultures could not communicate with each other and develop simultaneously in similar climates as they did in Eurasia. In North America, it was very difficult to travel from the West Coast to the East without the help of transport animals. Besides the vast deserts and plains of the North, the Rocky Mountains form a barrier.

Airplane travel has enabled humans to cross the Amazon forests only recently. With boats, one could only move along the river and not penetrate into the jungle nor cross the country. The Andes, like the Rockies, have also been a hindrance to communication between populations in the East and West. Apart from that, if prehistoric man wanted to travel from north to south, he had to endure frightening climatic changes. One can dehydrate easily a few hundred kilometers to the south, whereas freezing to death could be a realistic danger a few hundred kilometers to the north. In Eurasia, prehistoric man could travel in similar climates for thousands of kilometers from China to Portugal. The southern and northern parts of the continent, in other words, the rainforests of India and Indonesia, as well as Siberia, have remained havens for endangered species and hence allowed many species to survive.

Civilizations evolved in centrally located, mild zones. The first archaeological evidence of human occupation, for the northern and southern zones, appears thousands of years later. Humans tried to cross from Siberia into the Americas eleven to seventeen thousand years ago, during a period of crisis in the last ice age. We have evidence that they have hunted mammoths in Spain, Siberia, and the Americas.

As one approaches the equatorial belt, climatic changes increase. The feeding habits of the hunter-gatherer groups vary in different climatic zones. Unless it is a life-threatening situation, no human being would go to the trouble of passing through these sort of dangerous zones, which were controlled by different hostile tribes. There was also the question of knowledge of the different berries and game to feed upon for a hunter-gatherer. In other words, when human beings have no food problem, or do not encounter any threat to their survival, they choose to remain in place, to live and consume in pleasure.

The remaining continent, Australia, is a small one compared with Eurasia, has a harsh climate, and is barren compared with others as far as fauna and flora is concerned. Eucalyptus (which the Aussies call the gum tree), kangaroo, and koala come to mind as Australian species. The boomerang was known to the ancient Egyptians as well, but reached the present time in Australia and the natives of Arizona. One needs to briefly remember the expedition of Burke and Wills in 1860 to visualize the hardships of settling on a new continent.

Robert O'Hara Burke, an Irishman with the aim of crossing the continent on the north-south axis, started his expedition on the twentieth of August 1860. He was accompanied by the German doctor Becker, surveyor Wills, and approximately twenty others. The expedition team consisted of twenty-six camels, twenty-one horses, a copper bathtub, a book, twenty-one tons of food, and weapons. They faced death many times and were saved by the Aborigines, who survived these uninhabitable desert zones for thousands of years without any equipment, horses, or camels. After Burke was foolish enough to shoot at the Aborigines, many of the party, left alone in the desert, died from famine and scurvy (an illness stemming from the lack of vitamin C) despite all their equipment and knowledge.

At first, the "white" Australian community celebrated this expedition, but soon, the facts surfaced. Though this expedition took place "recently," I use terms like "approximately" or "possibly." The expedition consisted of approximately twenty people, because in those days, the number of Aborigine guides was not noteworthy. This expedition is famous on the Australian continent and has altered the approach of Australians toward nature, as well as their education system.

The Australian desert could at least be partially manmade, as the Aboriginals still hunt with the help of setting fire to the bush.

~23~
The Flying Doctor

In the winter of 1973, on a snowy February day, I left Frankfurt with a German academic exchange scholarship from Bonn University. I arrived in Sydney on a magnificent late-summer day. I started as a trainee, first in the Children's Hospital in Sydney, and then I visited the Flying Doctors' Centre in Broken Hill, NSW. From this centre, we would fly in small planes to help farmers suffering emergency situations. There were also weekly routine flights to native reservations. On one of these routine visits, the doctor I was accompanying picked out a few patients whom he knew well, with minor problems, and told me to examine them. He told me he would be having tea with the reservation nurse and asked me to join them when done with my chores.

One of the patients was this illiterate Aborigine lady, who was approximately eighty-five years old according to the patient charts. After listening to her and examining her, I gave her medication, as I had already been instructed by the doctor. Then it occurred to me to ask her a few questions about the first Australian farmers and sheep herders, as I realized I was sitting across from someone who had witnessed the settlement of Australia. When I asked her to tell me about those days, the lady looked at me calmly and said, "Well, dear, we didn't have any contact with them. They used to shoot at us, you know." I don't even put an exclamation mark at the end of this quotation, so soft spoken and matter of fact was her voice.

In Australia, the first European arrivals, on their first contact with the Aboriginals, thought that they had found the missing link between the ape and human beings. But in our vast planet, which is a biological laboratory, we have since learned that the natives of Australia are the first humans to make use of advanced stone tools. With no plants and animals to domesticate, naturally no large populations that would influence each other had formed on this continent. This racist view of the first settlers has proven to be wrong for the most basic biological reasons. This vast continent had been isolated until the 1800s. Archaeologists and paleontologists surprise us every day with new finds. On the other hand, New Guinea, which is only ninety kilometers north of the continent, owing to its tropical forests, different islands, and communication with mainland China, seems to be more developed.

Jarred Diamond, in his book *Guns, Germs, and Steel*, makes a good point when he states that a boy whose ancestors have lived as hunter-gatherers for thousands of years can become a jet pilot within one generation. Humans are very similar to each other when given equal opportunities.

~24~
Water and Global Warming

The first human civilizations, concentration of populations, and the birth of new ideas have taken place along rivers and main water streams like Tigris and Euphrates, the Nile, Huang He, He Chang, Indus, and a few thousand years later, along the Mississippi, areas where the biological diversity is large and where there is plenty of freshwater. We see the same story happening over and over, increasing populations that nourish themselves with an increasing carbohydrate diet. Salinization of soils, deforestation, and erosion follow. Human society is confronted with a serious problem of freshwater sources because of its increasing population and its impact on the environment.

A migration becomes necessary, accompanied by a longing for the "good old days."

Man seems oblivious to the threat to his existence as he makes lavish use of existing freshwater supplies and brutally kills off species. If you had to fetch your daily water from a well or river, how much water would you use?

Humans think that they can solve their water problem with the drip system, or mini sprinklers, or deep well pumps. But all they do is postpone the inevitable. The frightening consumption habits of Northern Europe and America, today's supposed civilized societies, but also of the rich and "successful" everywhere, are also causing major harm to the air we breathe.

~25~
Carnivorous Rabbits

Even to the naked eye, the amount of rich agricultural soils in the Americas, compared to the Middle East and central Asia, is striking. Of course, as always in biology, there are also deserts in the Americas, but the quality and the quantity of good soil compared with other continents is obvious. This friendliness of nature is also seen in Chile and Argentina, now the "better developed" South American countries. The elder cultures of South America developed in today's Mexico and Peru, which have similar natures to today's Near East. In Asia and the Middle East, people have simply devastated much of the soils that feed them over thousands of years. Environmental consciousness has been rising in the last decades, and this is a reason for hope as well.

The victory of the Western world over communism was partly because of the equalitarian ideal of communism and largely because of the fact that the United States, the flagship of the Western world, had vast agricultural lands only used for a few centuries. It was a less used new world. It is easier to indulge in pseudohumanism with a full belly.

At the end of the 1990s, after giving it serious thought in my room at the Desert Institute in Ashgabad, Turkmenistan, I gave up what I had named the Karakum International Project (KIP). I thought of the project after seeing the great Karakum Canal, or Karakumsky Kanal, which is the largest irrigation and water supply canal in the world. Started in 1954, it is almost fourteen hundred kilometers long, going through what has become four different countries and carrying water from the Amu-Darya River across the Karakum Desert in Turkmenistan. There was a massive loss of water along the canal due to poor construction and the salinization of the soil, causing an environmental disaster. (Salinization of soils is a phenomenon that was encountered in the Great Plains in the United States because of the lack of general knowledge of the farmer's world.) I had thought, "Why not make what you are doing on your own orchard here on a greater scale?," and started working on the project.

The Desert Institute was one of the most prestigious institutions of the Soviet Academy of Sciences. The pre- and post-Iron Curtain director, Professor Babaev, was enthusiastic and very supportive. I had been developing the plan and looking for international funds, inviting international scientists and discussing the chances of realization of the project. With drip irrigation, one could build up a vegetation belt along a portion of the canal and regulate the temperature in greenhouses with natural gas, as Turkmenistan is one of the largest suppliers of the world. One could grow carnations, sell them in Holland, and pay the credits and even make money on top of it. A twenty-five acre glass greenhouse lay abandoned and devastated near Asghabat, for example. It was a multimillion dollar project and the Turkmen State would guarantee the credits. But Turkmenistan was an unpredictable dictatorship that was living in total collapse in the aftermath of communism. Besides having to deal with a corrupt bureaucracy (that is, a classic kleptocracy), one had to deal in an environment of dubious interpretation of laws. When the big hotels were taken over by the Mafia, I reminded myself that I only have one life to live. It seemed much more realistic to engage myself with my orchard in the Aegean region. Karakum Kanal is a giant project, then and now not widely known to the Western public.

In these countries, one recognizes how irresponsible, thoughtless, and ignorant engineers of the Soviet Empire have been against nature. For years, they thought they had solved their problems with a simple engineering approach and too little biology! But of course the main idea of communism was wrong from the biological point of view: the idea that all men are equal!

Is there a contradiction in saying on the one hand that there is no noteworthy difference between human races, and on the other hand calling communism a biologically wrong egalitarian doctrine and saying that all men are not equal? From the biological point of view, there is no contradiction. Of course, there are cultural distinctions between defined districts, formed over thousands of years. But we can never be sure which setup will produce the next genius. In other words, it is the individual in a unique setting, not the race, that is important. The individual and how his "software" has been formed by a multitude of inputs are the significant factors. You never know from which brain the cure for your mother's illness will come.

Let us talk briefly about the evolution of civilizations and the formation and fall of cities, keeping in mind water and air (the most important factors for our survival). We have made mistakes, but can we avoid them in the future?

"Kleptocracy has increased as agriculture developed," said Diamond in *Guns, Germs, and Steel.* I fully agree with this conclusion. You hunt and gather and I steal it from you: This is equivalent to the idea that the goods produced belong to the state and you get your share of them. One needs a different class of people for bookkeeping and handling of the produced goods. However, politicians are human beings and have human deviations. We have to accept that bureaucracy is a necessity. Jealousy and nepotism are part of our nature. One who deals with honey will lick his fingers. It's normal that people in charge of anything will first think of people near and dear to them. It will never be possible to stop nepotism, but we can try to minimize it by being transparent and with "checks and balances." Humans always act in the same manner when they have unlimited power. They will abuse it. One has seen examples of this repeatedly in history in societies. One can observe it daily in the behavior of professionals, families, or relationships. I believe we simply have to see misuse of unchecked power as a part of the definition of our species.

As cities were established, rulers started searching for immortality with the help of their good-natured bureaucrats. They built great tombs, first for themselves and then for their ancestors. Finally, they erected their own statues. The statues probably came first, as those helped them solidify their power, their lives, and their egos. Only then did they move on to arranging a great death (so as to live on in history). Public sentiment turned against these statues, which were symbols of unrestricted power. And rulers who did not care about the public reaction continued erecting greater statues, buildings, and gigantic monuments. After a while, these became depersonalized as symbols of a belief, system, or nation.

We saw such symbols in abundance in Egypt, Anatolia, the Easter Islands, the Soviet Empire, and, finally, in Iraq. The symbols represent a widespread striving toward the comfort of one almighty and entail a class of people propagating, protecting, and benefiting from that almighty. This again is not good or bad; it is our nature. Is the family not a biological herd of mutual interest? Is the neighborhood something else? Is the family, the idea of tribes, then nations, really something else? Political parties, religions, governments, philosophies, and even mobsters are birds of a feather who flock together (gangs) for the common good.

Until modern times, powerful rulers who were not satisfied with these kinds of symbols divided people into classes and saw themselves entitled to spend the first night with any girl of their own people. This was called prima nocte. Taking the pretty women was one of the reasons/results of conquest.

The leader of the pack has priority in spreading his genes. We saw the rudimentary example of alpha-animal behavior in the Monica Lewinski case. Basically, despite the hullabaloo, deep down everyone understood the character of the alpha male and the episode went down as an amusing chapter in history. Many found it profoundly immoral, yet they had to hear every detail of the affair and could not help asking, "Where else did he touch you, Monica?"

Some early settlements got bigger and bigger and formed cities with need of wood, food, and water from the environment. Many would say it was the need for protection that caused cities to grow. It became harder to obtain these resources as the environment became barren. A cold winter, a dry summer, or an earthquake, flood, or epidemic was all that was needed to end a civilization. Cities were abandoned. Let us look at a scientifically documented example:

Anasazi natives (Navajos living in the area refer to them as the "ancient ones") constructed a 650-room, five-story building in the state of New Mexico more than a thousand years ago. That was the tallest structure built in America until the modern skyscrapers. The ruins of the building are now in the middle of the desert. What caused them to abandon such a big settlement? Archaeology and paleoarchaeology can give a clear answer to this question. At the beginning, two hundred thousand trees were needed just to construct the roof. Trees were cut from nearby forests. Then more trees were cut for cooking fires. The civilization devoured the surrounding forests.

As the countryside became barren, productivity in the surrounding acreage diminished because of droughts. And it got more difficult for men to carry wood for heating purposes. Finally, the Anasazis found it impossible to live in the desert they themselves created. Paleobotany can clearly document the date of the forest's destruction. U.S. scientists arrived at a similar conclusion for the city of Petra in Jordan.

Remember the island of St. Matthew? Do you think the only acceptable proof is for reindeer? If you do an Internet search for the island of Lisianski close to Hawaii, you'll read about rabbits that turned the island into a desert and then became carnivorous animals and ate their offspring. And then please look at Easter Island and observe a similar situation there, but involving man and mice this time! They cut down all the trees, and when the Conquistadores came, microbes did the rest. Other islands of doom are Norfolk, Henderson, Pitcairn, and Nihue, which were inhabited in prehistoric times by humans. After becoming barren and overused, they were abandoned.

You can guess how people abandoned the towns around the pyramids, Hatuscha, Ani, Dara, Pergamon, Aphrodisias, and Ephesus. For example, after earthquakes Ephesus's harbor silted up and now the city is far from the sea.

So Ephesus lost its entire economy, which was based on sea trade. Population increases, difficult infrastructure problems, problems with long-distance transport of food and fuel, an epidemic disease, an earthquake, or flood could have resulted in people's deserting the area. With humans' short memories, after a few generations, only tales and legends remain, even from times before writing was invented.

Even an amoeba strives toward light and warmth. All living beings try to optimize their positions. Humans are, of course, no exception. We "sense" what is good for us. So we are by definition interest-guided beings, racing each other if there is a conflict of interest. If our interest is in togetherness, as it often is, we will flock together in pairs or in families, as the smallest-interest guided flock. We have and still flock together in tribes, gangs, villages, townships, and nations. Within the nations or between nations are our interests in trade, professional groups, and belief systems, so in a sense we still function in gangs or tribes.

In a tribe of a hundred members, everybody knows everyone else. In a crowd of a thousand, that knowledge remains at the acquaintance level. Among ten thousand people, it is impossible to know everyone, but kinship can be determined after a short conversation. It's even possible to establish kinship within a million people, but beyond that, people can only classify themselves as from the same lineage. As the population grows, class differences become noticeable through clothes, hair styles, jewelry, mobile telephones, and cars. Even in a country like the United States, a person can be asked where he or she is from to establish acquaintance on a state basis.

People usually rate each other as being of a higher or a lower standing, and communication is based on this bias. If one approaches a higher-standing or more powerful person (like a general, a teacher, or a minister), communication will develop differently than it would if one had approached someone whom he or she believed to be a dependant or from a lower caste, like the cleaning lady in a hotel. A cultural reflection of trying to classify someone in the pecking order may be disastrous for communication if people can't approach each other without fear or feelings of superiority or inferiority. It is, on the other hand, normal for any institution or successful non-governmental organization (NGO), or even a functioning family, to be organized hierarchically.

~26~
Noah's Ark

During the ice age, so much water was captured in the ice that geologists' findings show that the sea levels were more than 120 meters lower than they are today. In the last few thousand years, they started rising a few meters. A few more meters and you have to abandon cities like Amsterdam, Rotterdam, New York, and London, to name a few.

The Mediterranean shores are the cradle of civilization. During the ice age, some parts of the Mediterranean Sea are reported to have been plains. As the ice age ended, the waters of the Atlantic Ocean flowed through the Gibraltar Strait to fill the Mediterranean area. Skeletons of pygmy hippopotamuses and mammoths found in five-million-year old rocks in Cyprus are evidence of this. There is a boat in every burial site of the ancient Egyptian emperors or kings, probably a direct cultural reminiscence of the great floods.

Approximately eleven thousand years ago, sea water began filling the Aegean plains, and then the Marmara and Black Sea "lakes" through the Dardanelles and later the Bosporus Straits. William Ryan and Walter Pitman caused a great sensation when they shared their findings with the world of science in the 1990s, and this increased the number of people searching for Noah's Ark in Eastern Anatolia. Ryan and Pittman tied this geological event to Noah's flood. Their theory made it possible to explain the legend of Gilgamesh as well as the later stories in the Old Testament. We see that the great "flood" that has shaped all our cultures was known to the Sumerians and Egyptians and later "copied" in the Old Testament, this time interpreted by the Jews, Christians, and Muslims as the wrath of God. It is easily understandable that the class of clergymen needed an easily understandable moral tool to lead the "flock," some sincerely believing in what they were saying and some even killing people who dared not to believe in what they believed.

Some 7,500 years ago, a great rush of sea water started flowing into the Black Sea, which had previously been a lake. Even today the salt ratio in the Mediterranean Sea is high. It goes down in the Marmara Sea and is at its lowest in the Black Sea. The 2.2 percent salt ratio in the Black Sea goes up in the Marmara and Aegean Seas and reaches 3.3 percent in the Mediterranean Sea. Evidence is accumulating that points to civilizations at the bottom of the sea along the coast of the Black Sea.

William Ryan and Walter Pitman interpreted the "rain" mentioned in the Bible as the melting of the ice age in geologic history and put a quotation from the Old Testament at the beginning of their book *Noah's Flood*: "And God said to Noah 'I've determined to make an end to all flesh, for the earth is filled with violence because of them . . . For my part, I am going to bring a flood of waters on the earth.' . . . The waters swelled so mightily that all the high mountains under heaven were covered for one hundred and fifty days."

~27~
The Middle Ages and Religious Wars

After roughly touching on the Sumerian, Egyptian, Hittite, and Greek civilizations from the biological point of view, let us now look at the Middle Ages.

In the Middle Ages, people learned about lost Greek civilizations from old Greek books in Arab libraries. The Berbers conquered the Iberian Peninsula between 712 and 718; in 730, they began retreating during the Reconquista Wars and left their mark on the peninsula for about seven hundred years. In 1492, the "Reconquista" of Granada was achieved.

With its four-thousand-year written history, China developed in a mostly uniform fashion because of its close proximity to the West, with the Great Wall and the mountains on one side and the sea on the other. For example, the West learned of gunpowder, porcelain, compasses, paper, and the kite from China. Porcelain-making was known in China as far back as the first century BCE and was perfected in eight hundred years, while Europe only unlocked the secret of porcelain-making during the reign of King August der Starke (August the Strong) of Leipzig in the 1700s. After that, porcelain manufacturing started in Sevres, France, and in England by Darwin's grandfather Wedgewood. An invention's journey in the opposite direction took place when microchips were discovered in the United States and traveled to Japan, Korea, and China.

In the Middle Ages, kingdoms and the Church lived in mutual tolerance. The Church dealt with science and religious matters, while feudal rulers, aristocrats, and kings took care of worldly matters.

Most probably, immigration from the Eurasian plains began in the fourth century as a result of those plains turning into deserts. First the Huns, then Seljuks, and finally the Ottomans came west. Between the years 1095 and 1271, during the Crusades, some European kingdoms moved toward the East for a period.

In the years from 1299 to 1918, under Ottoman rule, a more or less stable administration existed in the area.

~28~
The Armenian Kingdom and the Ottomans

In 1986, while visiting the ruined city of Ani, in modern Turkey, just across the border from what was then the USSR, I read on an inscribed panel that the king had gone into the forest and cried after a major earthquake. Right across from the Arpaçay brook at a distance of 100 to 150 meters, Soviet soldiers were keeping watch.

I wondered where the king had found a forest to cry in. The nearest forest was one hundred kilometers away. The Forest Administration had posted large wooden signs along the highway proclaiming "Forest Is Life," "Trees Are Your Friends," "Love the Forest and Protect It." Within a rocky steppe of tens of kilometers, the only wood one could see were these signs. To be frank, some fascist ideas came to my mind in connection with the bureaucrats who had put so much wood in the middle of a rock desert. I bitterly wondered what they were thinking as they used wood to make these signs! Probably they did not think at all. They must have received orders from their central administration to place these slogans along the road and they must have thought of making them using the most beautiful and cheapest wood (what is cheaper than wood if you are an officer of the Forest Administration?). I had thought, "If I were a poor person living in such a place, where the temperature goes far below freezing for at least four or five months every year, the first thing I would do would be to burn these wooden signs."

Goats are the livelihood of most of the villagers in the region and goats eat the newly planted trees! It isn't easy and requires insight and commitment to solve the environmental problems of which we as humans are a substantial part.

The architect of the cathedral of Ani also repaired Hagia Sophia in 889. Then this city was conquered in 1064 by the Seljuks, who built mosques, madrasas, and Turkish baths. It was abandoned three hundred years ago after an earthquake. When Istanbul had a population of three hundred thousand, it is surmised that Ani's population was one hundred thousand.

We often read about the tragic events that took place between the Turks and the Armenians during the First World War, as the Ottoman empire was disintegrating. This is even reflected in the relations between modern Armenia and modern Turkey. Allow me to share a personal story:

My niece is an attractive young lady in her thirties; she rides horses and motorcycles elegantly and was selected for the national horseback-riding team to compete in hurdles. Because of her love for animals, she became a veterinarian. She then received training to become a race horse trainer, and won the Gentleman Jockey's race as a jockey in 2006. Her grandmother, my aunt, in other words, twice went on the pilgrimage to Mecca. My niece fell in love with a young Turk of Armenian origin in Istanbul and became a Christian in 1999, as this was the wish of the groom's family and the young couple did not have strong feelings about religions but wanted a nice church wedding. In January 2000, we all celebrated their wedding at the St. Mary Armenian Orthodox Church in Istanbul. Solutions for such "intercultural" problems are always the same. As a person who believes in the selfish human gene and also that humans can live peacefully in the world, I hope all those involved will understand that harmony between Ireland and England, Spain and Basques, or Israel and Palestine is much more to their benefit than is conflict. Even when people believed in a heavenly justice, they acted with vengeance and a payback mentality over generations of evolution as they instinctively realized that the fear of payback was the only system of checks and balances they had. This was carried over for generations, the biased approach constantly contaminating the future.

Let me ask this simple question: Who are the Turks? Today's understanding is that 99 percent of the people living in Turkey are Moslem. Let's take a bird's eye view of the people living in this region and their beliefs:

It is not known how many thousands, or tens of thousands, of Turks came to the West from central Asia. What is known is that they came into contact with monotheistic religions around the Caspian Sea. Hugh Pope, in his book *The Rise of the Turkic World*, tells of Uighurs in the Chinese province of Xinjiasng-Urumchi protesting and chanting the name of an exiled leader named "Isa." Isa means "Jesus" and is today a common first name in Anatolia. Also Musa, or Moses, is not unusual as a first name. The brother of the founder of the Seldcukhian Dynasty, Selchuk Han, is named "Israel." So if the other brother had founded the empire, it would have had a different name. We don't need to outwit ourselves when researching history. Some things are to be taken at face value, nothing more. Linguistically, the difference between the Turkish spoken in Turkey and Azerbaijan does not differ more from one another than, say, the English spoken in England and that spoken in Texas. The difference becomes more pronounced when we go farther east. Compare the languages a western Turk speaks with the Turkish spoken in the Chinese province Urumchi, or with the language spoken in Uzbekistan, comparable to Dutch and German spoken in Southern Germany, for example.

In regard to the names Constantinople (meaning Constantine's city) and Istanbul, some Turks think that people who use the name Constantinople over six hundred years after Emperor Mehmet II Fatih conquered the city are biased. Some of these people do use this name spitefully. Just some information to people who have emotions: The name in use now, Istanbul, is older than Constantinople and is also derived from the Greek words "Ins tan Polin," meaning "into the town." I sincerely fail to understand what difference it would make if this name was derived from a

Sanskrit or a Chinese word or an expression in an African dialect. People will go on teasing each other and making politics out of everything; it simply is our biology.

The number of Turks who migrated to Anatolia is unknown. What is known is that they were introduced to Western religions while they were living around the Caspian Sea and that a majority chose the Moslem religion while some of them became Christians through a series of deals. It is thought that Ashkenazi Jews immigrated to Europe from the shores of the Caspian Sea. So that region was a cultural center of importance then. The same sort of bargaining took place later in England, during the time of Henry VIII, between the Catholics and Protestants, the British ending up with their own brand of Christianity.

The Russians became Christians during the reign of Vladimir the Great, the grand prince of Kiev in the year 988. He had decided to become monotheistic and sent envoys to study the various Western religions. Vladimir consulted with Jewish envoys (who may or may not have been Khazars), saying that their loss of Jerusalem was evidence of their having been abandoned by God. To Muslims, their religion was undesirable because of its ban against alcohol and pork. In the churches of the Catholics, his emissaries saw no beauty; but at Constantinople, where the full festival ritual of the Byzantine Church was set in motion to impress them, they found their ideal. If Vladimir was impressed by this account of his envoys, he was yet more so by political gains of the Byzantine alliance. Most people who believe in religions think they belong to a religion because it is TRUE, but they have no idea that they are being manipulated by forces that seek power through mind control over and over throughout history.

The Ottoman sultans thought themselves to be descendants of the Oghuz lineage, a tribe coming west from central Asia. The ridiculous aspect of the lineage concept can be easily seen in simple family-tree research: Ladies who bore sultans were either of Circassian, Greek, Romanian, Bulgarian, Albanian, or Bosnian origins ("sultan" is a word for emperor or empress). Of thirty-six sultans, only the mothers of the first two were "Turks." Is it important to know how much Bulgarian, Greek, Hittite, Kurdish, Arab, Bosnian, Armenian, Circassian, Italian, or Turkish blood runs through the veins of Turks living in Turkey today?

After conquering Constantinople in 1453, the Ottoman Empire, which reigned in Greece, Bulgaria, Romania, Montenegro, today's Hungary, Albania, and parts of Austria until the nineteenth century, was defeated by the Habsburgs in the Vienna War of 1683. This was the beginning of the regression period for the Ottomans. Just before the First World War, they were the "sick man" on the Bosporus.

The most important element for the Ottomans' success in their attacks and movement west was neither their ability to make better use of gunpowder or cannons nor their being more athletic, clever, or better riders and warriors. It was rather that they offered an alternative to ambitious people living in "Old Europe," where a caste system prevailed and where it would be impossible for a hungry and driven individual to reach a higher status within a period of ten to fifteen years. A serf was doomed to remain a serf, though a eunuch in the Ottoman court was also doomed (once a eunuch, always a eunuch, I guess). But a harem was only for a few, most of the Anatolians being able to afford only one wife!

For a slave or a peasant tied to his feudal ruler by prima nocte, Ottomans were a breath of fresh air, as they were for an ambitious craftsman and those who felt themselves under the pressure of "religion." There was not much choice in the European Middle Ages for those who did not belong to the aristocracy.

In contrast, in the Ottoman Empire, only the sultan's family belonged to the aristocracy. There was no other aristocratic class. Those who were successful either became pashas (generals) or were appointed as bey (governor) to some region, no matter what their cultural or geographic origin happened to be. Their children didn't inherit any title. Appointees who had received their assignments without racial discrimination could stay at their posts as long as they were successful. Failure could mean exile or death.

While the peasants spoke Turkish, Ottoman administrators spoke Ottoman, which was a mixture of Arabic, Persian, and Turkish. The same thing was also seen in the Russian Empire. The governing class found it more elegant to speak French. It was the same in many small European kingdoms. Scientists also set themselves apart from the populace (herd) by speaking Latin.

The fellow in charge in the decentralized system had all the authority and could decide and govern in a pragmatic way, but had to answer to the emperor. The systematic insecurity of the appointee was later a method practiced in all democracies, the United States being at the top of the list, contrary to the inflexible centralist Chinese systems or later that of the Soviet Empire.

Aside from immigration and the rich farmlands at the root of the United States's scientific superiority lie elements of nature and competition. Many scientists compete with each other, but only a small minority can gain the title of distinguished professor, represented by "tenure." "Old Europe" has only now started to move away slowly from the established science bureaucracy in the last few years, copying the United States. European systems worked fine and produced results in the eighteenth and nineteenth centuries, when many assistants were ready to work without pay when the world population was 700 million to 1.5 billion.

~29~
Religious Wars Again
and Reformation

After its war with the Arabs in Iberia, Europe hoped to find salvation in many of its old values. Tens of thousands of Jews fleeing the Inquisition found shelter in the Ottoman Empire. Even today, it is possible to see traces of past history in the technical and political collaboration between Israel and Turkey. Few in the West realize that Turkey, a Moslem country, trades and has a close relationship with Israel.

It is reported that only one-third of the population of Southern Germany survived the Thirty Years War between Catholics and Protestants (1618 to 1648). It would not be wrong to assume that more than one-fourth of the total population of today's Germany and of neighboring countries perished through hunger and epidemics as a result of these wars. In Germany, in the wars in 1524 known as the "Peasant War," villagers rebelled against religious and aristocratic pressure.

The development was not very different between the sects in Islam, the Sunnis, if you wish, comparable with the Catholics, and the Shiites comparable with the Protestants. Each country has slightly different versions of belief systems, like in Western Europe, and some belief systems coexist altogether within a region. Whereas during the reign of the Ottoman emperor (sultan) Selim I (1465–1520), there was consolidation in the Moslem world. Especially in the Anatolian Alevite (Alaouite) sect of the Shii, there are ideas reminiscent of Zarathustra and ideas rather free of dogmas. During the reign of this emperor, who was also the caliph (the religious ruler), differences between the sects were assumed to be nonexistent. After the Turks conquered Istanbul, a cultural war began in the Catholic world and a place was opened for new interpretations by eradicating taboos and sacred cows.

Western Europe's influence, which can still be felt in today's world, was achieved as a result of the Renaissance (thirteenth to sixteenth centuries), with the emergence of scientists like Newton and Darwin, partly with the technological revolution helped along by climate conditions (a colder climate zone with the habit-forming need to plan for winter), and partly because of religious wars and learning to live together. The natural resources gained from water mills and coal mines in England and Germany played a meaningful role as well.

~30~
Tobacco and Coffee

First found in America, the spread of tobacco in the then-stable Ottoman Empire was typical. The myth of Turkish coffee and Turkish tobacco took off. Because of their strong and orderly system, Ottomans focused their search on their questionable pleasures enthusiastically. Western Europe was living a polyphonic life because it was made up of small kingdoms, cultivating new colonies. The Ottoman colonies in Arabia were the producers of coffee.

While countries in the south (both in South America and Southern Europe, like Italy, Portugal, Spain, Greece, the former Yugoslavia, and Turkey) lost ground in the nineteenth and twentieth centuries from languor mostly caused by climate, the North, with its planning habits acquired during long winters and partly with the use of hydraulic energy, realized the Industrial Revolution. Coal was once today's crude oil and the motor to development!

In my opinion, the greatest change as far as ideas were concerned took place within the framework of the Thirty Years War and the Reformation. People who differed from one another grasped the necessity of living together and tolerating (bearing) one another. They realized that not doing so would be very costly.

Farther south, the Arab nations, following their advanced civilizations, remained under the rule of the Ottoman Empire, then under British, French, and Italian rule in the nineteenth and twentieth centuries. But with nature's blessing of petroleum, they seem to have achieved a peculiar progress in the twentieth century. When one looks at the way Arab countries progress, one cannot help wondering whether it is a blessing or a curse to have petroleum. Long before oil was found, though, the Arabs were stagnant. Their "flowering" was during the Middle Ages, while Europe was in the Dark Ages. And what since then? Plantations that once grew coffee beans used in Yemen and Mocha coffee have been deserted. Like everybody else, Arabs now buy their coffee from South America.

In the twentieth century, two communities, one in the Mediterranean and the other in a similar hot climate, showed different patterns of development: Australia, with a population mostly consisting of immigrants from northern Europe, and Israel, with immigrants from all over the world interested in having their own country. For mankind, immigrations are the most effective elements, next to the natural ones, causing cultural changes (paradigm shifts).

~31~
Discovery of America

The World Is Flat by Thomas L. Friedman, an inspiring narrative of the twenty-first century, starts with an entry from Christopher Columbus's diary in 1492:

"Your Highnesses, as Catholic Christians, and princess who love and promote the holy Christian faith. And are enemies of the doctrine of Mahomet . . ."

Not much more need be added; the entry is self-explanatory.

We know, on the other hand, that Columbus contacted many European monarchs and Moslem rulers before he persuaded the queen and king of Spain to bankroll his search for the riches of "India." If he had found the "venture capital" he needed in the Ottoman Empire, how would his journal entry have read? This is just one of the many examples of interest-guided understanding and application and nothing more. A proverb nicely captures such loyalty based upon self-interest: "I will work with the sword of the Lord whose bread I eat." Closer to home there is the expression, "He who pays the piper calls the tune."

Closed communities could not develop further to surmount their rigid caste systems. India, China, and Japan started slowly profiting from encounters with different cultures during the 1800s. Ottomans, having swallowed a big chunk of the world with Constantinople, continued their victories for a few centuries but were more absorbed in the joys of life, like coffee and tobacco and harems. Until the 1800s, the Ottomans were a world power. Smaller, more competitive, and hungry kingdoms in the West started acquiring colonies and industrializing. The Ottomans slept through the Industrial Revolution, becoming the "sick man" on the Bosporus.

Pause briefly and think about how world history might have developed if the Americas had been "discovered" under the flag of a Moslem or Chinese emperor. It was the development of new technologies and the need of small, competitive European kingdoms for new colonies (Lebensraum) that brought a new era in human history. The discovery of fertile lands began dictating the future of the herds. The natural resources produced well-fed people to lead human accomplishments. As these were largely of the "Christian-Western" background, this became the leading culture.

We have seen how, throughout history, contacts with different cultures have fostered innovation. What is true for human cultures is also true for all living creatures, including plants, with which we share a lot more biologically than many believe. In fact, immigration has been the pacemaker of innovations. A clash with different cultures has always been fertile. The question is, Do we have to pay as much in human lives for it in the future or are we capable of learning from the past? Americans were lucky enough to learn to live with different cultures as it became a huge melting pot.

In botany, we can perhaps see the need for a multicultural habitat more easily than in human settlements. The more uniform the landscape, the more natural balances are distorted and the more fragile nature becomes. It is true that everyone sees the importance of biodiversity, but not everyone sees that human diversity is a good thing. I mean, blacks moving in next door often upsets white Americans. I suppose a large group of Arabs or Americans or Chinese, for that matter, buying up houses and moving to Urla wouldn't enthuse my neighbors either.

In areas where monoculture is practiced (where only one type of crop is grown), the natural balance of nature is destroyed and the land becomes unstable. Agriculture in itself is a monoculture of some sort. This happened in Mesopotamia, where over-usage and deforestation diminished the soil, in the Middle Asia plains where forests perished, and finally in the great plains of the United States, where corn caused the problem.

Under these conditions, the soil is subject to erosion, and soil formed over thousands of years is washed away by water and wind. Or because of drought, salting, or depletion of soil nutrients, there is failure in the one crop that is grown. This can result in famine, as happened in Ireland between 1845 and 1849, when the potato harvest failed. As a result of the famine, approximately two million hungry and impoverished Irish people migrated to other countries.

Just as in biology, monoculture, with its blind spots and prejudices, causes catastrophe in communication among men. The same pattern can be seen. Every culture has its own prejudices. If we want the earth to be ruled by a monoculture, we risk not having any place to which we can migrate. We don't see things as they are; we see them as we are, or as we would like to believe them to be.

It is impossible to know in advance what species will be useful to mankind. Therefore, human beings show a rightful reaction to the one-type state or one-type trade cultures ruling the world, even though administrators and military and big businessmen might like everything to be uniform and easy to control.

An innovation not allowed in one culture could prove itself in another in a few decades. An idea that is not understood in one culture can be hailed in another. Is Columbus not the best example? As they age, humans understand that they can learn a lot from history (if they reread it) and from experience. This is why I feel longevity is a hope for our race, because if you live longer, you have more time to observe and to think about what you observe. This is a controversial topic, though. The youth of the world call us "geezers" and do not think our great age gives us much insight! In fact, they might think that the reason most politicians and political leaders do such a poor job is because they are old. It is neither a cultural matter nor a problem of age, but of an educated and intelligent minority versus the rest. This minority of people who are lucky enough to have both education and the intelligence to understand trends have an obligation to the rest.

~32~
America's First Discovery

The first larger groups of human beings apparently came to the Americas between eleven and eighteen thousand years before Columbus. I don't want to go into further scientific detail of the dates, as it is not the purpose of this book, but I do want to share part of an e-mail I received from American Archaeology Society member Dr. Curt Hoffman on March 31, 2006, when I wanted to obtain more information on this subject:

"Well, you've stepped into the stream of a major controversy in archaeology! I attended a whole conference about this in South Carolina last fall. There is indeed a growing body of evidence that supports a pre-Clovis occupation of the New World, at least as early as 20,000 YA and possibly as early as 50,000 YA." (YA: years ago.)

Around thirteen thousand years ago we find lots of so-called Clovis settlements (named after the first settlement found in Clovis, Mexico), and another simultaneous finding is outstanding: the disappearance of mammoths, giant squirrels, and other animals. One can easily visualize the bloodbath people made. It is understandable that people killed a mammoth and lesser known species to obtain a few tons of meat, and then eventually moved on. They would probably (not having deepfreezes or ever having felt the need for them, nor understanding the "luxury" they lived in) have let the rest simply rot at the hunting site. Scientists guess that the disappearance of horses and camels in America took place exactly this way.

Man had a significant effect on nature long before writing was discovered and we could document what we have hunted to extinction. On the Galapagos Islands, the animals do not run away from people as a natural act, as they had not seen a man until the discovery of the islands. For an animal to develop and inherit the fear of man, it would require thousands of years of evolution.

We are doing the same thing today in terms of our treatment of such resources as air and water, which we seem to believe are unlimited. If you think that enough drinking water is not a problem, you should realize that this is only the case for the lucky half of mankind now, and that a serious problem for the whole world is expected beginning in 2025. You see that our species is unbelievably indifferent. Aren't we an incredibly dumb species? Man relaxes in the misleading comfort of the thought, "After me, deluge." Human beings, due to their instincts, do not take any action until the last minute, and sometimes when the knife cuts the bone they find themselves unable to do anything. We have laziness in common with our roaming relatives and also with many others that need a winter sleep.

Whether we like it or not, we'll see in this century whether man has learned anything from history. In a 1991 interview, Jacques-Yves Cousteau said that one U.S. citizen causes more harm to the world than twenty Bangladesh citizens. These figures still hold. He said that population was the world's most important problem and that he considered high-consumption communities as a mechanism eating itself—as a cancer. He was talking about stopping population increase in fifteen years. In my opinion, he was not mistaken; we ruin hundreds of species every year by encroaching on their environment, and at this time many species in the world are fast disappearing. We have no chance other than stopping population increases and finding a natural balance if we want to stay on this island. Few Americans have any idea that their lifestyle is destructive. And all those Turks driving huge SUVs around the narrow streets of Istanbul are just the same.

~33~
Garden of Eden to
Paradise Lost

The overkill theory probably provides the best explanation for the extinction of mammoths worldwide, not only in the Americas.

In the El Pindal Cave in Spain, it is possible to see a mammoth drawn by "primitive people." Today, even a well-educated and talented artist could not draw a mammoth more beautifully. What captures one's attention is that the heart is in the right place. Approximately eleven thousand years before Harvey discovered "blood circulation," this was drawn to facilitate hunting and to explain that if the spear struck that part of the animal, it would be easier to kill it.

Paul Martin, the Darwin of mass extinction theory, describes in very persuasive language how the four-ton ground sloths in America or Moa birds in New Zealand were hunted by man until they were extinct. This, of course, is only a theory, since no one actually observed the vanishing of a species due to hunting eleven thousand or one million years ago. Even if someone who saw this wrote about it, who would believe the document? Especially with the disinformation campaigns during wars, one gets confused as to what to believe. But when findings of unbiased scientists start to overlap, theories receive general acceptance. It is only natural that we will get contradicting and confusing data concerning everything at any point. But when we look from a distance, things will become clear and the trends become obvious. Bias can have many origins, oft due to scientists' craving for fame or due to cultural bias giving priority to the work of scientists from a particular culture. The principle of one hand washing the other! After studying for several decades, you can oft tell who is pushing knowledge and in which direction.

If we take into consideration the amount of time it took for Darwin's ideas to be included in school books, in spite of the increasingly fast information transfer, it will take several years yet for these findings to become a part of the general culture. On the other hand, while there are beautiful girls in bikinis and football and basketball scores in our lives, why would people be interested in a paleontologist finding mammoth dung or bones in a cave?

Martin says the following: "It would be absurd to assign blame to the progeny of Paleolithic Europeans or of the first Americans for the extinction of the Old World or New World mammoths, to Australian Aborigines for the end of diprotodonts, or the New Zealand Maoris for eliminating Moa. It is important to remember that the extinctions of the near time occurred worldwide. To the extent that responsibility is assigned, it belongs to our species as a whole. This may be an even more disturbing thought for many." Is this not also true for the wars we had and all the bloodshed in our history?

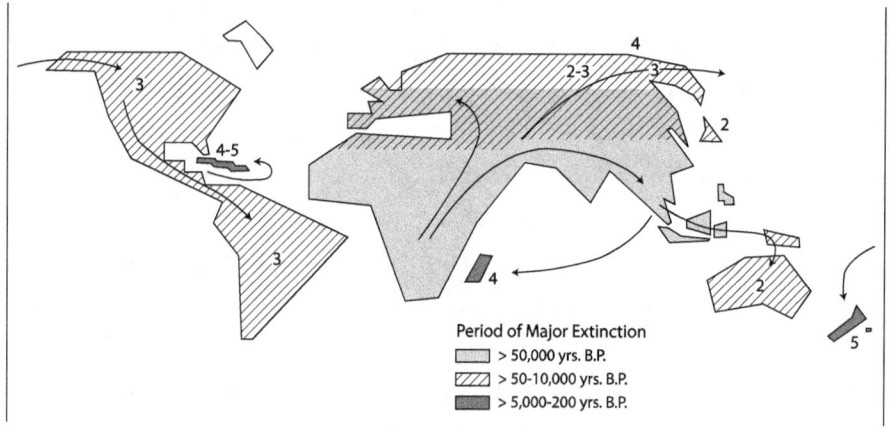

Fig. 3: Sequence of Extinctions

Arrows indicate direction of human dispersals: Numbers indicate order of human settlement. (Adapted from Martin, *1970 American Institute of Biological Sciences: Twilight of the Mammoths,* Paul S. Martin University of California Press, 2005.)

Diamond suggests that "maybe like the Thanksgiving day in the memory of the first European settlers one should have a Thanksgiving day where very first Americans cherish the mammoths who fed the humans and became extinct in the process." This is not only the case for the Americas, but also for the whole world, as these animals fed the ancestors of Germans, Russians, Chinese, and Mongolians as well. Findings suggest that the last mammoth died in Siberia at the time of the Pharaohs. I think of human history roughly in three periods:

I: Ice Age & Floods (the age of mammoths and leather shoes, the wooden age or prehistory)

II: Ancient Egyptians, Sumerians, and Hittites and the Classic Period (ancient Greeks, Arabs, and Romans) through the Middle Ages until the Americas' Rediscovery (the age of horses, the wheel, pottery, and writing)

III: The Present Time (engines, airplanes, communication, and the computer age)

PART THREE
DOCTORS, LAWYERS, RELIGIONS, AND MONEY

~34~
Physicians and Diseases

We have seen how the human population has grown in the approximately two hundred thousand years since Homo sapiens has been around. Let's look at physicians and diseases, going back five to six thousand years in an attempt to understand biology's influence on history.

Physicians, like lawyers, are usually close to those in power, and the same can usually be said of clergymen as well. We can be sure of one thing: Any statesman who has consolidated his power (by election or overthrow) and whose business has found its groove, in whatever century, will show and demand nepotism. In addition, to a certain extent, this nepotism will be tolerated and expected by the ruled. We will dislike a politician who favors his relatives directly and obviously (unless we ourselves are the relatives), but indirect fringe benefits are always there. Why else would one support a politician if one didn't hope something positive for oneself (whether it be a favor when the time comes, more wealth, or better protection)?

So let's stick to reality; it's so much more exciting than imaginary tales. Looking at what we have learned in the light of scientific research, disease existed even before humans. This can be seen in the bones found in fossils. Just as pathologists today can map a corpse's medical history, paleontologists can determine how prehistoric animals died by examining their bones.

As humans evolved from a hunter-gatherer society into a sedentary one with the help of domestic animals and farming, the disease rate increased significantly. At the same time, because larger populations lived in close proximity, they were able to exchange ideas and develop career specializations. As societies grew, they needed individuals to take on different trades and jobs. This division of labor led to increasingly complex technological, political, and economic systems.

~35~
Civilization: A Curse?

For example, an interesting finding is that according to their fossils, Anatolian hunter-gatherers who lived eleven thousand years ago were taller than the Anatolians of the farming societies and kingdoms thousands of years later. Only now are today's Turks, Greeks, and people in the Balkans almost as tall as their hunter-gatherer ancestors. The current height of these populations seems attributable to advances in medicine, nutrition, and communication. The fact that settling down in communities and becoming farmers caused a decrease in size makes us question the value of cultural history.

Anthropological finds show there was a difference between the height of the ruling classes and the ruled, as can be observed in some societies today. The hunter-gatherers surely had a class structure as well as a method of worship, a clergy, and burial ceremonies. Their paintings and carvings were, as we can see from examples excavated in Çatalhöyük or Göbeklitepe (11,500 years BP), or in the French or Spanish caves, as good as any modern artist can create. However, the paintings became fixed and stereotypical when humans became sedentary. Within hierarchies, there were restrictions on what the disciples of the earlier artists could draw. One sees in Egyptian and Hittite drawings a stylistic control and a lack of free expression. All the figures look alike. Humans started drawing alike as they learned craftsmanship from each other. In the centuries and millennia that followed, the paintings and drawings became more picture-like; artwork was flat, two-dimensional, and lacked a feeling of motion. Actually, it was not until the last century, when "moving" pictures were discovered, that art once again became as free as it had been in the Stone Age.

When the population of a settlement grew large, tribal law, with one chief making all decisions, was no longer sufficient. Rulers realized the need for writing and the importance of codifying a set of written laws.

The earliest examples that have survived are the laws of Hammurabi, the king of Babylon, and those of the Hittites. These ancient legal codes show us that as far as justice goes, not as much as we would like to believe has changed in the intervening centuries.

Hammurabi laws (1782 BCE) state the following:

215. If a physician makes a large incision with an operating knife and cures it, or if he opens a tumor (over the eye) with an operating knife, and saves the eye, he shall receive ten shekels in money.

216. If the patient is a freed man, he receives five shekels.

217. If he is the slave of someone, his owner shall give the physician two shekels.

218. If a physician makes a large incision with the operating knife, and kills the patient, or opens a tumor with the operating knife and cuts out the eye, his hands shall be cut off.

219. If a physician makes a large incision in the slave of a freed man, and kills him, he shall replace the slave with another slave.

220. If he opens a tumor with the operating knife, and puts out the patient's eye, he shall pay half his value.

221. If a physician heals the broken bone or diseased soft part of a man, the patient shall pay the physician five shekels in money.

222. If he is a freed man, he shall pay three shekels.

223. If he is a slave, his owner shall pay the physician two shekels.

224. If a physician performs a serious operation on an ass or an ox, and cures it, the owner shall pay the surgeon one-sixth of a shekel as a fee.

225. If he performs a serious operation on an ass or ox, and kills it, he shall pay the owner one-fourth of its value.

~36~
Hammurabi and the
Hittite Laws

We still have vengeance and punishment as foundations of justice and international law. Contemporaneous Hittite laws took approaches similar to those specified by Hammurabi. As the components of writing are the shared inheritance of humankind and pass from one tribe to another with minor changes, it is likely that laws among the Assyrian, Babylonian, Hittite, and Egyptian civilizations influenced each other and were understood and applied similarly. In Hittite law, however, the principle of an "eye for an eye" was abandoned in favor of "compensation for an eye." There are records showing that even the act of murder could sometimes be financially compensated for rather than simply punished. Furthermore, Hittite legal codes imply that a crime is an individual act and only the criminal is to be punished. This punitive-to-restorative legal approach still has to receive popular acceptance throughout the world. An important change in modern justice will be realized if the principle of reparation for a crime, rather than revenge against the criminal, is implemented. In the United States, such schemes are known as "victim offender reconciliation programs"; in Britain, they are called "reparation schemes." Integrating a philosophy of reparation into our educational system could help prevent tragedies like the famous Columbine shooting and the massacre at Virginia Tech. A lot of research is being done in this area, and I find it especially amusing that some scientists try to seek the roots of restorative justice within the Christian religion. (Didn't the Hittites predate Christianity by about two thousand years?)

We see that in the last few hundred years slavery has been gradually abolished. The freedom of any human being and his right to pursue happiness is more or less the agreed norm today, but what about equality?

It does seem that if you have a certain passport, you are more equal than those who do not hold such an equality document. We receive daily bulletins of the numbers of Americans dying in Iraq. But what about the Iraqis? Why do we hear so little about their deaths? Unfortunately, if you compare the life stories of the victims, apart from their different passports, you will find that they had similar lives; they are mostly poor people. Unlike what some would want you to believe, most U.S. soldiers in Iraq are not fighting for a principle they believe in; the majority enlisted in the military because they wanted a secure job, a livelihood. The matter is as simple as that; a "volunteer" army is really a mercenary army.

We all know that there are laws for the rich and laws for the poor. O.J Simpson's trial was a showcase for all. Why else should people want to be rich? Okay, what do you do after you are rich enough to be on the safe side? Do you belong to the rich who feel that being rich is all that matters, or do you seek to be one of the rich, compassionate, and intelligent people who sense that real security can only be achieved in a better-educated and better-fed world?

Our system of justice is still not much further developed than when societies were ready to hunt and prosecute the children, or even the grandchildren, of a culprit. For example, why are the Germans still paying for the horror of their grandparents under the regime of the fanatical lunatic Adolf Hitler if this is not the case? They have paid billions of dollars in reparations to the descendants of the Jews killed in concentration camps (to be exact, 63.2 billion Euros through 2005, and one must realize that a large portion of this sum was paid some thirty years ago, when money was worth a lot more!). Some lawsuits involving Nazi crimes and forced work done in the Third Reich are still ongoing. The foolishness of disparaging a race has already been proven on the world stage. Today, Israel is gaining the admiration of the world by its contributions to humanity in numerous scientific congresses. The main motivation behind this is simply the communal solidarity brought about by thousands of years of exclusion from other societies, as well as cultural discipline and the will to work. Since they branched out around the world, Jews have benefited from close encounters with all kinds of civilizations and cultures, learning from their accumulated knowledge and experiences. The Jews have never had the luxury of fencing themselves off from the exterior world in an unchanging cultural atmosphere, unlike the Chinese.

The entire world has realized that although they have been exiled to various places throughout history, Jews have not ended up as an inferior, cowardly people with a second-rate culture. John J. Mearsheimer and Stephen M. Walt, two U.S. political scientists, analyzed the role of the Israel lobby on U.S. foreign policy. The State of Israel has laws prohibiting a Palestinian who marries an Israeli from becoming an Israeli citizen. It is ironic that the state founded by people who suffered the most from racism became a state of people whose attitude to those around them is racist.

Human cultures have produced today's justice and democracy as the two most important (yes, vital) institutions. Democracy is another sacred cow we need to put under the microscope. Although we admire democracy and generally agree that it is preferable to most other systems, we should remember that Hitler was elected to the Reichstag and was later made chancellor of Germany. How many German voters knew he would become a mass murderer when they elected him? Many of his supporters just wanted jobs and a strong leader. As Churchill said, "Democracy is the worst form of government except all the others which have been tried."

Let's return to the history of medicine. Have we reached a point where a man is paid to cut us open and is only rarely penalized for not producing a successful outcome? There is no clear-cut solution to the balancing act between helping people and the need for surgeons on the one side and greed and abuse of power on the other. In medical history, religions forbade physicians to work with cadavers because "a dead body is God's property." There's not much difference among religions on this subject, although some writers like to add a sentence against "the other religion" under the guise of science.

In order to acquire scientific knowledge beyond that available to the general population, physicians are known to have taken risks. Was this done for the sake of science alone? Was it to acquire fame and fortune? Probably some of each, depending on the individual.

~37~
A British Trade in Corpses

Examination of patients became routine only in the last one hundred years. In 1898, Sir Arthur Conan Doyle, a medical doctor better known as the creator of the famous detective Sherlock Holmes, recorded the terror of a woman who would not allow her chest to be examined: "Young doctors allow themselves such liberties, my dear." Patients weren't examined in the past because of modesty, especially women patients. And doctors didn't know much about examination of patients either, as they weren't allowed to dissect corpses because of religious restrictions.

In China, physicians carried with them a woman's figurine, and the female patient would indicate the location of her own pain by pointing to a spot on the figurine.

Autopsies on corpses were performed by gutsy physicians after they overcame the "body belongs to God" prohibition. The development of knowledge of anatomy and surgery wasn't easy. Physicians were tried in court for performing autopsies as recently as two hundred years ago.

When it became obvious that autopsies were helpful in understanding disease, a booming market for fresh corpses developed in England. In Scotland, a gang "producing" fresh corpses was arrested. In the well-documented case of Burke and Hare, the two were found to have killed sixteen people and sold them to the anatomist Robert Knox. Justice made a deal with one and hung the other. Knox was set free, but was a broken man thereafter.

This is just a historical example showing once again what a rotten lot we are. Man murders for the flimsiest of reasons. All over the world, I've witnessed deaths caused by physician error being swept under the carpet. I feel sorry for that ambitious Scottish doctor who so craved knowledge about anatomy that he did not care where his "experimental" materials were obtained. The greed of the murderers supplying him with fresh corpses hasn't changed either. You can still find people who are willing to kill others for money. We can read in any day's newspaper about murders similar to those two hundred or two thousand years ago.

There are also plenty of people today willing to kill out of preconceived cultural assumptions in revenge for a crime they believe was committed against their ancestors. If you can't do something positive to mitigate their need for vengeance (and I can't do anything positive either) who can? Surely not those corpse dealers.

When I am forced to conduct some bureaucratic business or enter into an unnecessary and unreasonable argument, I get a fresh understanding of how humans can be driven to kill each other. Man carries the sum of his evolution within himself and can revert to his most primitive stage of development at any moment. It has taken thousands of years for us to understand this simple fact.

Those close to the ruling class generally became physicians (in other words, alchemists). Barbers, dentists, and blacksmiths worked with tools and so became surgeons. A word about professional deformation: Any doctor, teacher, preacher, police officer, businessperson, soldier, farmer, or bureaucrat will be molded by his professional life. A doctor who is used to seeing ailing patients will eventually start treating everyone with an attitude of superiority. A teacher who mostly deals with students who usually know less than he does will start thinking that all people know less.

After a while, people with a fixed salary will be defined by whatever profession they are in. If you are interested in seeing professional deformation, it is easy to see the difference in the worldview of an Air Force officer and an Army officer, or even between an artillery gunner and an infantryman.

Other examples of professional deformation can be seen in the way a bureaucrat or a politician deals with the state's money. Because the money is not his own, it is only something abstract for such a person. A bureaucrat is tied in a chain of command and responsibilities and can't easily act differently, even if he wanted to. He is neither good nor bad. He will naturally fit into the system showing the least resistance. What we end up seeing as kleptocracy in most cases is only the way good people try to survive in any larger system.

Businessmen usually tend to avoid problems and lose interest when there is a risk or an investment that demands too much involvement. Some will eventually take bribes for granted, while others will never come into contact with bribes. Everyone will learn the shortcuts each profession offers. I myself have never encountered a lawyer or a judge with any deformation of any kind. On the other hand, even if I had, I should be wise enough, especially about the judges, to let the reader figure that out.

To this day, if you go to a hospital in England, you will find two physicians on night duty: the internist Dr. Smith and the surgeon Mr. Jones. A surgeon is addressed as Mister. In this way, the past remains alive. The herd instinct comes out victorious here as well. Surgeons reach a certain professional deformation and aim for cutting and pasting, whereas internists go for the more conservative approach, that is, they lean toward the status quo and slow progress. Surgeons often fall victim to unnecessary hustle and bustle and may take actions that cannot be undone.

On the subject of income, physicians can reconcile their differences just like members of any other profession. But what has become obvious is the rivalry and disparate approaches to medicine among internists and surgeons. Without any doubt, this interdisciplinary competition is an engine that has driven medical progress and has been the impetus for many advances in medicine.

With the advent of endoscopic surgery, which non-surgeons can also perform, internists have become more like surgeons. Again, the desire to amass more income was an engine of progress. You can always charge more if you can have an anesthetized patient.

Professional people are more defined by pecking orders. Uneducated and poor people tend to have more common sense than many educated, successful people with obvious professional or lifestyle deformations. Pecking orders, or hierarchies, and the spaces of freedom, or personal spheres, are the definition of human cultures. Common sense is the invisible ruling order of the free spaces.

~38~
Pergamon and Ephesus

Not far from where I live is the ancient Greek city of Pergamon (the word "parchment" comes from this name), probably the most important center of medicine when the world population was around 250 million. Ephesus is an hour's drive away. On my orchard, also an hour and a half away, are Lydian burial mounds.

Perhaps the reader will not recognize the word "Lydian." Croesus, king of Lydia, was once known as the world's richest man. He minted golden coins from mines and from the sands of the river Pactolus. Treasures stolen from his capital, Sardis, were purchased by the Metropolitan Museum in New York and only returned to Sardis through the efforts of a journalist and the local museum director. Then they were stolen again in 2006, this time from the local museum. The town Urla, where I live, has the oldest known prehistoric harbor (Limantepe) built by man, dating back to some forty-five hundred years ago. Next to it is the oldest known settlement in this region, dating back some six thousand years.

My geographic proximity to all these ancient sites naturally doesn't qualify me to offer discourse on history, but hopefully their accessibility should not prevent me from learning and thinking about the past.

It would not be wrong to label the legendary Hippocrates (470 to 410 BCE) as "the father of medicine." He was probably born on the island of Cos and practiced in Pergamon. To him is attributed the quote, "Those who want to perform surgery should go to the battlefield."

Galen (131 to 201 CE), a native of Pergamon who practiced in Rome some five hundred years after Hippocrates, is known as the Western father of surgery. He became a leader in anatomy and in cutting and sewing wounds while treating gladiators and wounded soldiers, a fortuitous circumstance that allowed him to look inside that closed box, the "sacred" human body (as ordained by the Church and by the cultures before Christianity). An even older father of surgery was the Indian Susruta, who is thought to have lived and practiced on the banks of the river Ganges in 600 BCE.

Even the ancient Greeks, and especially Aristotle (384 to 322 BCE), had looked at the beating heart of a chick and defined it as the punctum vivendi, the living point, or the "heart of the matter," as we say today.

~39~
William Harvey (1578 to 1657)

William Harvey's "scientific definition" of the circulation of blood opened a new era and changed our lives, perhaps as much as Newton's definition of gravity. Today, every child grows up with this knowledge. Understanding of these two concepts is so vitally important that one wonders why people didn't discover them earlier.

Untold numbers of humans were sacrificed for thousands of years. In addition, animals have always been killed for food. Arms and legs have been cut off in battles for thousands of years. Why then wasn't blood circulation noticed and realized before Harvey? Why did it take so long to accept this reality?

The answer is simple: We are a dumb and rather lazy species. We will respect sacred cows, rules, regulations, and established power, with inevitable taboos, whenever we can and keep our mouths shut for the sake of our insignificant daily advantage, or what we believe is to our advantage. Be it for knowledge, be it for justice, or be it for protecting our "way of life," we will lie (or simply remain silent) when we sense a personal advantage. The people in the establishment are happy with what they know and do not really want to know about fancy new ideas, which could distract them from their routines (though they will of course pay lip service to the idea of being open to new ideas). Our capacity for deceiving ourselves and our obstinacy in not admitting faults is unbelievable. A German proverb says that dumbness and pride are carved from the same wood. And in which society has speaking up not harmed the whistle-blower?

People seeking knowledge are easily stopped by the jealousy of others. "This is neither the right time nor the right place" for this or that topic has always been a killing phrase, sometimes burying a good idea for ages. The human need for freedom of thought has survived with the help of sarcasm, wit, and humor.

These are thoughts I could share with an eighteen-year-old shepherd of average intelligence.

~40~
Why Are Humans
So Slow to Understand?

Wouldn't this be a smooth transition to the subject of prejudice and common sense? Just a little patience, please. We're getting there. A primary school graduate with some practical life experience (our shepherd, for example) will understand and listen, but will not see how it will help with problems associated with births in his flock or how he'll find the money to buy a better ram. Nor will he be able to understand how this kind of learned talk will make an impact on society.

He would start losing interest in these things because he wouldn't understand how our conversation could affect the subject of Catholic-Protestant bloodshed in Ireland or who should control Jerusalem. If the shepherd had had just a little more education and opportunity, he might be able to check the facts outlined in this book. Maybe in ten to twenty years he could reach the level of defending these ideas. So what is the difference between you and me and the shepherd, if we presume we are of similar intelligence levels? Could it be only our formal education of between ten and twenty years and our highly enhanced self-esteem? I believe that the "differences" between societies are not any greater than this because today's mass communication methods bring knowledge to everyone!

If the problem is reduced to ten or twenty years of education per individual, plus the self-esteem and experience, maybe we can work toward a solution. With today's level of communication, we might be able to reach a consensus within a few hundred years, if we are realistic. But it may be even simpler. There is a common sense in all humans apart from all personal and cultural bias, greed, and egoism. People who have obtained a certain level of education and culture have started to have fewer children; in fact, in the developed world, the population growth rate is falling.

Unfortunately, a decrease in population is contrary to many government policies. Yet, if this general consensus to have fewer, or even no, children among better-educated, wealthier humans is not a reason to be optimistic, I don't really know what is. This is the common denominator we have been talking about and a far better starting point than many other solutions to world problems. As they have recognized the diminishing world resources and applied those consequences to their personal lives, more and more people have started bearing fewer children as a personal choice. People who don't have the means to understand what's happening to the world are continuing to breed as dictated by their instincts. But there the lag is also getting shorter; until a hundred years ago, only a few people had the means to travel around the world and had the education or the insight into different cultures to get a comparative understanding of death. These would be successful people in their late forties or fifties, and we know that there were only a few of them. People would have had few encounters with death if they were not in the medical profession. A class that taught them about death was needed. Today, more and more people in their twenties are getting a realistic understanding of death and of other cultures thanks to television.

People go through both active-creative and passive-slumber phases in their lives. In any case, 30 percent of an average day consists of sleeping (or being unable to sleep); sometimes 90 percent is taken up by snoozing and eating. The more mechanized our daily routine becomes, the less activity we engage in. These active-passive periods of our lives are like a daily stroll through different stages of evolution.

People follow routines for decades, falling into the vicious cycles of "I'm going to die anyway," "après moi, le deluge" ("after me, the deluge"), or "What do I care?" By nature, humans relax with mild to strong addictions to routine or fun or intoxicants. Tobacco, alcohol, over-eating, cocaine, opium, and marijuana are the indulgences found in periods that sometimes last for decades in a man's life, or hundreds of years in the life of societies. The current generation is going through a period of television, alcohol addiction, and depression.

But if a man feels he's been threatened, unfairly treated, or humiliated in a difference of opinion or a simple scuffle, then this animal's blood pressure rises. A level of hostility, epitomized by the phrase "if looks could kill," is reached quickly. Moreover, this feeling can be shared with others, even over longer periods of time, and in some cases over generations. Most people spend much more time dealing with other people than with abstract ideas. Feelings like hatred, jealousy, prejudices, and ambition are topics well known to almost every intelligent human.

Every act of communication, in the final analysis, is an act to change something, and this impetus for change can be seen as an act of hostility or aggression. As long as this aggression doesn't lead to physical violence, we will learn to tolerate and live with it.

The situation is no different for communities. Sometimes cultures undergo dormant periods lasting for hundreds, or even thousands, of years when they are unable to acknowledge realities that have been under their noses all along. Then, when a new way, a new discovery, is made, the public implements this change with a herd mentality. Or for some imaginary reason or because they are unable to rebound from an argument, they adopt an adversarial attitude to the innovation and this attitude may last for decades. The herd mentality can only be overcome through dialogue.

When we look at history, we see that it takes quite a long time for an idea to mature. Newton, Harvey, and Darwin are some examples given in this book. It is not important whether Tom, Dick, Harry, Charlotte, Bao Chan, or Piraye puts forth a certain idea. It takes time for these ideas to mature and become widely accepted. Of course, people without

Newton's scientific definition of the principles involved had been utilizing the force of gravity all along, even before domesticating animals.

There were 250 million people on the earth during Christ's lifetime. The majority were ignorant and their life expectancy was limited to thirty or forty years. When we consider today's 6.5 billion population, it could theoretically originate sixty more Jesuses, fourteen more Shakespeares, and twelve more Darwins. Obviously, a defining genius can be born in any time period and in any population.

We have a lot of simple, helpful realities. We have travel and communication patterns like never before. Although the world is becoming arid, the ozone layer is quickly getting thinner, and seasonal changes and aging problems are debilitating us, there is no need for us to be pessimistic. We don't know exactly how global warming will affect us in the future, but we can surmise that it will initiate many mutations in a great number of species. What we do know with certainty is that the world population will age considerably and an age bomb will hit us within the next few decades. Could it be that within this fact lies a hope for the future? Our world is so mobile that, for the first time in history, we now have millions of people who are aware of different cultures, have worked in different countries, and know that people aren't so different wherever they live.

The spread of progress takes place according to the "monkey see, monkey do" principle. When a discovery is made that answers a need, people and cultures copy each other, sometimes adding their own innovations. This copying is neither good nor bad. It is merely a biologically given pattern, which functions at unconscious, subconscious, or conscious levels in all living beings.

By working with the unbiased facts we already know and looking at these facts from different perspectives, we eventually come up with new solutions. These innovations may be no different from what has already been said. But as more and more people reach an age at which they can grasp new ideas and more and more people have the power or the means of communication to share these ideas, there is hope that it will make a difference.

When ideas are explained clearly in mathematical or scientific terms, they become very difficult to reject, even in an atmosphere of chaos and prejudice. They become reproducible and are thus a foundation for further ideas. That is why Harvey's definition of blood pressure was a stepping stone for modern medicine, as it was based on simple, repeatable numerical measurements that everyone could verify. I hope that the number of people who know enough about the negative effects of overpopulation has reached a critical mass so that we can now strive toward a solution to our environmental problems.

~41~
Infectious Diseases

Medicine has advanced by trial and error. The Black Plague, or Bubonic Plague, is believed to have originated in Egypt in 540 CE. It hit Istanbul two years later, devastated the city, and kept spreading to kill an estimated 25 percent of Europe's population. That corresponds to more than seventy-five million people today.

The "discovery" of the Americas was an exchange of diseases and illnesses. When Cortez conquered what is now Mexico City with approximately three hundred men, three hundred thousand citizens died of smallpox, not from war. And Pizarro, with his band of 168 Spaniards, conquered the Inca Empire (which had an army of eighty thousand men) with the combination of better weapons, horses, and, of course, European diseases. And the result of this conquest? That would be the humiliation and huge inferiority complex that is readily seen in Native American cultures today.

The reaction of people to cultural and personal humiliation, which they take to be communal injustice, has always been strong. The first phase of humiliation in these cases is the physical defeat of war. The second phase is the cultural despair of those defeated, and the last phase is the self-humiliation that sets in.

It is interesting that even today Cortez's native mistress is called a "traitor" by some in Mexico because she served as an interpreter for him. Still, after five hundred years, the victors are the "others" to some native Mexicans. And this attitude will remain if no one makes the effort to inform people that wars and infectious diseases are purely a historical reality. If we keep on reliving supposed historical and pathological injustices committed against our ancestors, we could lose our chance for survival on this planet. Whether it is Hitler in recent times, or the Japanese-Chinese war, or the conquest of Constantinople, or the invasions of the Americas, the same pattern is seen again and again.

When violent conflict occurs and the reason for the conflict cannot be understood by all, the seeds of further violence are sewn. What we can call our cultural heritage will have a tendency to emerge whenever the losing side feels that the time is ripe for payback. When an epidemic occurs at the same time as the conflict, such as during America's discovery, or during World War I, then the degree of perceived humiliation by the vanquished can seem even greater.

~42~
Indians, Bosnians, Conflicts

Following Tito's death, Yugoslavia disintegrated. By the end of the 1990s, some Serbs under the leadership of Milosevic were systematically exterminating the Moslem minority. I recall the words of my high school classmate Claus Lutterbeck, then the Vienna correspondent for the renowned German journal *Der Stern*: "They are killing each other," he said. The tone of surprise in his voice still rings in my ears: "They are full of hate as if the Battle of Amsfeld which occurred in 1389 took place yesterday."

We know what followed. Serbian leader Milosevic, who died in prison in 2006, was forced to leave the presidency in 2003 after a multinational military intervention led by the United States. We know that one party's killing of another is never a solution. How will violence ever result in peace? For instance, how will the strife between Israel and the Arab world, or that between India and Pakistan, ever be resolved? Let us go back to the new center of the world, the Americas. The Indians of Peru still think of the Spanish Conquest with contempt. What is the common denominator between Indians and Bosnians?

Conflicts arise wherever there are people. Generally speaking, humans' encounter with nature is also full of bias and hence conflicts. In our relationship with nature, so far it has always been nature that was wasted.

If groups can be separated by belief, race, or language and define themselves as "self" and "other," a foundation for future problems is established. More and more, racial and religious differences are becoming secondary, and the primary reason for conflict between peoples is that of perceived economic injustice. The Moslems of Arab origin in France, and especially in Paris, revolted in 2006. The Watts Revolt (the black-white conflict) that took place in Los Angeles in 1965 was basically for the same reason.

Let's look at a cup that is filled daily with drops of prejudice and humiliation. These drops can overflow or freeze and cause the cup to break. Here race and religion help the flock to form and recognize one another. The gist of the matter is one of growing injustice and the pessimism of individuals who do not see a way out. They resent the dominating people or cultures who, in turn, compound the misunderstanding by acting superior to the ones under their domination. This is unfortunately in the genetic makeup of humans. The less fortunate, when they cannot find a solution, will endure it. But in our modern world they can easily communicate with each other, as we saw in the peaceful protest of one million "illegal" immigrants in May 2006 in Los Angeles. These were people with hopes. If hopelessness is predominant, conflicts will increasingly erupt into violent protests.

If people begin to feel that death is preferable to their living conditions, then those who think that they have come to a dead end concerning their "race" or "culture" can turn themselves into live bombs.

Strife between husband and wife belongs in the same category. In personal encounters, a quarrel may start whenever people think that they are being unjustly treated. However, it takes more than feelings for a whole group to feel discriminated against. Characteristics like race or religion can easily help to create or define a group. Our selfish genes are our only hope for survival on this planet. The better we understand that equal opportunity will ease such tensions, the less we will have such conflicts disguised as religious or racial outbreaks. People have been killing each other over boundaries for thousands of years. In the fight for more personal space, people in masses are slowly realizing that having fewer children leads to more personal space, while having more children causes boundary disputes. This is a selfish and personal response, but it is a reaction that shows hope for mankind. If more families adopt this selfish approach and see the world as their home, then most of the world's problems will be easier to understand and solve.

Wars either end in the victory of one side over the other or when both sides exhaust their energy and are therefore unable to actively continue the conflict. When our selfish genes take over, and in the light of today's knowledge, more people will understand that it is against their interests to fight. When differences of opinion become conflicts, it is always cheaper to search for a compromise. After all, life is a compromise and justice is an agreement that people can only reach after they communicate.

Even when there isn't a war, bickering continues at all levels. Sometimes this habit becomes downright funny. This belonging/not belonging or self/non-self dichotomy is often defined by the district where one lives, not by religion or by race: People who commute to New York City every day for work are called the "bridge and tunnel crowd." They, in turn, question the sanity of noncommuters who spend a fortune on small apartments in Manhattan yet fear to go for a walk at night.

In personal conflicts, it is often more fruitful when a neutral third party intervenes. We must somehow make more use of this moderation when the third party's interests are minimal. Compromise is heavenly justice!

In conflicts between nations, if the mediator is another nation, then the third-party interests sometimes burden the mediation. A look at an illness could help in understanding the necessity for a different approach.

The word "hysteria" comes from "hysteron," which means uterus in Greek, and came to acquire its current meaning by its use in explaining behavioral disorders brought on by menstrual problems. The illness called hysteria is seen in southern societies more than in northern ones. To a psychiatrist, the term "hysteria" or "hysterical conversion" is used, for example, to describe when a man suddenly becomes paralyzed or loses his eyesight on finding his wife in bed with another man. This person actually cannot see or walk even though doctors cannot find any neural damage or physical changes in his brain. Such people can live the rest of their lives in their "hysterical" condition. They are not fishing for sympathy. Sometimes such illnesses can be cured by a "miraculous" treatment.

This hysterical reaction doesn't set in when the man who finds his wife in bed with another man reacts in a conscious practical level, such as, "Shall I kill him and go to jail?," "Oh, now I can divorce my wife without giving her 50 percent of my income and assets," "Shall I beat him up?," "Oh, my God! What a huge man. If I say something to him now, what if he beats me up?," or "Is this the man who is paying for this house?" Such analytical thinkers never present the clinical picture of hysteria.

Speaking in international terms, a sort of mass "hysterical conversion" reaction has been seen in the Australian Aboriginals and in Native American Indians after their contact with the "white man," but also in African nations and Asian nations after the Europeans were the ones to find the New World and use the resources there.

The other type of reaction, the desire to "punish" what is seen as an offence, is the typical vengeance reaction that ends in bombing a country or a region and killing people by the thousands. It will only cause more and more conflicts and violence and encourage more groups to seize the opportunity, setting up a chain reaction for yet more violence to occur. This is the predominant way we react today.

The third form of political reaction was demonstrated by Jesus, and later by Gandhi, and yet later by Martin Luther King Jr.: the decidedly nonviolent reaction. If this level of response is reached and accepted by a majority, then we will have hope for peace in this world. The current lawmaking and international politics are predominantly dictated upon us not by compassion, but another basic biological urge: vengeance.

~43~

Horses, Guns, Germs

In *Blood and Guts,* Roy Porter says, "Guns and germs together enabled tiny European forces to conquer half a continent." In *Guns, Germs, and Steel,* Jared Diamond deals with the same theme.

The ancestors of Incas who saw the Conquistadors coming on horseback had seen neither horses nor gunpowder nor smallpox nor measles nor typhus. They thought of guns as thunder and the men on horses as gods. They had not experienced any type of Eurasian flu. They died of these new diseases by the millions, so much so that indigenous peoples have almost become extinct in Argentina. In Chile, some remain. In many other countries in South and Central America, they form the majority.

Native Americans were isolated from the technology that had been developing in the Old World for thousands of years. It was with this technology (and their infectious diseases) that the West could conquer the Americas.

We know that cultural changes start within the embryo. Even if only the father smokes during his wife's pregnancy, the developing embryo becomes a passive smoker and may not attain the size and development standards of children of non-smokers until age eleven.

Cultures clearly differ in how infants are delivered (by midwives, in hospitals, by relatives), are cared for (carried on the back, turned over to nannies), and are raised (by both parents or only one parent, by relatives, in institutional settings). Further complicating matters, mothers and their children consume different food in different cultures and wear different clothes (or none at all), have different proximity to animals, hear different voices, and so on. It is therefore clear that differences that developed over thousands of years cannot be abolished overnight. Nor should they be. However, as equality of opportunity increases, such differences will become less important as conflict-causing factors. On the other hand, each individual is different; being slow, for example, can be a handicap here or a life-saving difference there.

We know that Columbus's success was due to the flexibility he showed by going to a variety of different sources to raise the funds for his journey. We as humans have begun to understand the importance of the right use of special plants and minerals in curing disease. A variety of approaches to medical treatment is beneficial. What is true of human thought is also true of plants.

What do the Conquistadores and diseases, and different germs and plants, have to do with culture, population, and the environment?

Perhaps the history of a species of plant in the genus ferula, a member of the carrot family, will help us understand the importance of a single plant. This plant could have accounted for some of the main themes of this book except for one factor—it became extinct during the age of the Roman Empire. It was used by Roman ladies as a contraceptive and, as the need for the plant and its price rose, it simply became extinct. Many a plant useful for human ailments and sought by the pharmaceutical industries faces the same fate today: an example is the devil's claw, or harpagophytum procumbens, used in the treatment of arthritis—an export from Namibia.

~44~
Inoculations and Beautiful
Caucasian Girls

Smallpox was, or is thought to have been, eradicated in 1979 with the successful vaccination campaign of the World Health Organization (WHO). Children are no longer vaccinated against smallpox. President Bush started vaccinating first-aid personnel after 9/11 as a precaution against potential biological warfare.

It is thought that the first vaccination came to the Ottomans from Caucasia, Circassia, and Chechnya, where midwives, by scratching the body and blowing a live virus on the wound, made the person immune to diseases. In the Ottoman era, Circassian girls were considered beautiful because they had no smallpox scars on their faces.

The wife of the British ambassador to Constantinople, Lady Mary Wortley Montagu (1689–1762), not only wrote home about the vaccination against smallpox, but also saw to it that her own children and two princesses were vaccinated in 1717 and 1722, respectively. Sources inform us that at that time 60 to 98 percent of children in England and Germany died of smallpox. According to some scientists, a virus is the smallest living entity, while according to others, it is not even alive, as it has no cell with a nucleus or its own metabolism. A virus thrives on its hosts, and the illnesses it causes come in epidemics, so those reports of extraordinarily high smallpox mortality rates do not sound unreasonable from a physician's point of view.

Voltaire, George Washington, and Benjamin Franklin, whose son died of smallpox, were involved with spreading the "Turkish" vaccination.

The Edward Jenner method now used in smallpox vaccination (he himself was vaccinated with the Turkish method) was a vaccination made of cow virus and had minimal side effects. The Royal Society did not want Jenner to give lectures on this method, so in 1778, Jenner published his seventy-five-page study on his own. In 1805, Napoleon ordered all the soldiers under his command to be vaccinated, but by 1874, the French had lost interest in vaccination.

In the Franco-Prussian War, while about 23,400 French soldiers were dying of smallpox, the Prussians lost only 294 soldiers. It is not unreasonable to say that smallpox vaccination was the single most effective factor in today's human population growth.

It would be inconsiderate to write on health and population and not mention Ignaz Philipp Semmelweis (1818–1865), a doctor who studied infections during childbirth. Through his research and practice, female life expectancy became longer than men's. Semmelweis noticed that fewer mothers died in childbirth when midwives rather than doctors tended to them. This, he felt, was because doctors operated in an environment with many diseases and did not wash their hands before delivering a child. Midwives, in contrast, operated in a more hygienic environment. But convincing doctors to wash their hands was like attacking another cultural sacred cow. It is not easy for anyone in a system to change things.

~45~
Cholera and Others

Cholera, a Sanskrit name, is most probably another disease of Asian origin that has caused millions of deaths. A person who has a normal immune system will not be affected by the disease, even if he drinks five cholera vibrion (as these bacteria are called) in a glass of water. Nothing will happen even if he drinks a thousand. It is only dangerous when there are more than thousands of such active germs (a critical mass) in his water. I have shared this information to point out how enduring humans are. Nevertheless, you should of course boil any water that may contain the cholera germ.

Malaria is another epidemic that continues to be a problem, and typhus still continues to be the cause of death from time to time.

The invention of antibiotics (as a result of a simple observation by Alexander Fleming that bacteria cultures in a petri dish don't thrive as well when there are fungi on the same dish) proved to be an important victory over bacterial diseases, or at least it was thought to be. Though the practices of circumcision and cutting holes in people's skulls are thousands of years old, surgery only entered a golden era in the 1950s, when new techniques in anesthesiology and antibiotics were applied. In the sixties, we used to believe that all microbes were our foes and antibiotics our savior, but science has since shown the flaws in that assumption. For example, the koala bear is a good example of how we need bacteria to survive: The koala eats eucalyptus leaves only. To digest these, they need some bacteria in their intestines. Baby koalas begin by eating the feces of their mother, and thereby gain the bacteria they need to digest eucalyptus.

In the antibiotic era, heart, kidney, and liver transplantations have become routine. Thinking of the first open abdominal surgery ever, performed conveniently in a graveyard in 1790, we can see that we've come a long way. This first explorative surgery was performed during the reign of Louis XVI, and he had permitted the surgeons to have a look inside a prison inmate who would otherwise have hanged anyway. After the explorative abdominal surgery, the surgeons left the poor man in the graveyard to die. To the astonishment of the grave diggers, the patient survived the night and was then set free by the king, the same ruler who was sent to the guillotine in 1793 after the French Revolution. It took about a hundred years until abdominal surgery was tackled again, this time with the aid of ether, chloroform, and curare.

Yes, whether in Bosnia or America, whether among people or between bacteria and people, tension accumulates and causes eruptions. Conflicts cannot be prevented by prohibitions. Neither can they be prevented by keeping silent, as Wittgenstein recommended. As previously mentioned, conflicts can only be overcome by talking and sharing and understanding the mechanisms: Love and knowledge grow by sharing.

We do not know how many millions of people will die of AIDS, a disease that spreads partly through the act of love. (Some estimate that there will be between seventy and one hundred million deaths from AIDS by 2025.) The knowledge that sixty million died of the Spanish flu over a period of a few years causes people to worry now about the possibility of an epidemic of bird influenza.

I would suggest that people who want to see the interaction of medicine and human life objectively and entertainingly should watch *Lorenzo's Oil*, starring Nick Nolte, Susan Sarandon, and Peter Ustinov. It is a realistic film about the conflicts and financial issues in the field of medicine today. In the film, the story of a father striving to cure his ailing son from a rare fat metabolism disorder is dramatically told.

In this film, we see how the world of medicine is run by money and also how money does not account for everything. We also see how an educated man who is not from the medical world succeeds in having an enormous influence on medicine by doing research and working relentlessly. Looking at the matter superficially, we can see that any young person who is a high school graduate can penetrate into the extreme boundaries of medicine within as short a period as two to three years. Here is just another reason to work hard and not give up.

Computer tomography (CT), magnetic resonance imaging (MRI), and extracorporeal shockwave lithotripsy (ESWL) are some modern methods of medicine with a tendency toward inflation. These techniques are mainly operated by profit-making medical centers. ESWL focuses energy in the form of water waves on a kidney stone in order to break it. The technique first appeared in medical annals as an interesting new development in the 1980s, and I was professionally involved in its early phases.

Just as ocean waves break up huge rocks without harming the sea, one can direct waves produced outside the body to break up kidney stones. A stone within the body is broken by this simple principle and is expelled through urination as dust or sand particles.

As an internist, I didn't like using stents—devices often used by urologists that run along the urinary tract and are inserted into the kidney endoscopically. We used stents in only 6 percent of cases and published this figure. In medical practice, the insertion of stents occurred in just over 30 percent of cases. By my own accounting, stents and accompanying overnight stays in the hospital annually increased health costs by more than ten million dollars yearly in Germany.

Should you be curious, ask your nearest ESWL center about the percentage of cases that receive stents. In the year 2000, when I migrated to the Aegean, there were six ESWL centers in both New York and London, cities comparable to Istanbul in population. In Istanbul, there were forty-four ESWL centers. These numbers clearly show the difference between "regulated" medicine and the early capitalist Wild West medicine where inflation is achieved, resulting in patients with no kidney stones being sent to ESWL for treatment for a referral fee. But on the other hand, this shows the greed of humans everywhere.

In Australia, at the beginning of the year 2000, there were so many radiologists involved in an MRI scandal that it seemed possible that new radiologists would have to be imported if all the ones involved in the scandal were punished. The scandal of so many radiologists being partners in MRI centers was the main topic in all the newspapers over the Christmas holidays. Nothing much happened, of course.

Referral fees are often a real part of the medical business. Though this is supposedly illegal and unethical in the United States, a U.S. weekly on July 2, 2006, ran the following caption: "Money Machine: Is Your Doctor Sending You for an MRI Because You Need One—Or Because He Needs the Cash?"

In 1982, in Lübeck, Germany, I met Dr. Friedrich Wegener, whose name was given to Wegener's granulomatosis, a disease of the lung and kidneys. As a retired gentleman, he was attending a symposium at the university. As a young man, he had not been granted the noble title of professor because he was not on good terms with his boss. He could have gone to other schools if he had found the title very important. He preferred to live in the beautiful city of Lübeck as a "doctor" only. Americans used to consider the title PhD as the highest academic rank and not use the title professor in areas other than at the university. They used to retire only as doctors, leaving all other titles in their offices, and this was a cause of admiration among the scientists in Europe for a long time. Now we see that U.S. scientists have started to like the title of professor at international congresses and in newspaper articles. As in other cases of inflation, when this title gets too much use, it will find its true value as well.

After I felt that I had made enough money in my profession, I invested in agriculture. I started to get interested in other fields and reading once again. I then thought that we doctors made up the biggest immoral gang worldwide—I still believe doctors are greedier than, say, soldiers or men of the cloth (in general, that is), but I don't believe my colleagues are the greediest lot anymore. The following joke might express my feelings: After a caring person like Mother Theresa died, she was given the option to walk through the four different levels of heaven and pick a level for herself, as she had done so much good during her lifetime. The first level she visited was inhabited largely by men and women of the cloth and other bureaucrats who had kept away from bribes and worked hard during their lifetime; on the second level, the majority consisted of honest women and men who had also worked hard during their lifetimes. On the third level were innovative philanthropic people who had also set up foundations with their earnings. When she came to the fourth level, she was astonished to see a lonesome and huge grizzly bear. When she asked why the grizzly bear was on the highest level of Heaven, Peter responded, "Well, he ate three lawyers when on Earth."

~46~
A Final Cure to Penis Cancer

Jews, Coptic Christians, Orthodox Christians in Ethiopia, and Moslems practice circumcision. In this practice, the foreskin of the penis is cut off. Among the city-dwelling Jews and some Moslems, this procedure is performed right after birth. Among many Moslems, this operation is usually performed at age eleven or twelve in celebration of the child entering puberty, as a kind of "bar mitzvah" ceremony in which the child is initiated into adulthood. This is a tradition dating back to the prophet Abraham.

As the ancient Greek story goes, when the fox with a short tail became king, the first order he gave was to have all the tails of the foxes in the forest cut to the same size as his own tail. I assure you, this is only meant as humorous discourse on our cultural evolution. I guess it has simply evolved as follows: Some six or seven thousand years ago, a slave who did manicures and cut wounds, much like those who do "piercings" in our day, cut the foreskin of the penis of a slave as they were playing around, and he liked it. In time, one such person with a circumcised penis became the tribal leader, and then circumcision became the fashion. Later, it became a part of culture. Some practiced it, and some did not, until some scientists in New York found that circumcision, to a certain extent, prevented penis cancer. Since then, circumcision has been carried out by some for "scientific" reasons. People naturally like to play with their penises. So other scientists started doubting the assumptions of the aforementioned scientists. It is true that it is easier to keep a circumcised penis clean, but today people who take regular showers can stay clean without being circumcised.

Or if you prefer a cynical approach: Penis cancer, which is rarely seen, will completely disappear if we cut off all penises at the root. By the way, eye cancer is also widespread, and if we take out an eye when a baby is born, I am sure that we can prove statistically that this cancer risk will also be minimized by 50 percent! The percentage of cataracts people have will also be cut in half. But this is of course an attempt to give the reader a humorous view on how statistics can be manipulated to prove almost anything and what some grown-up scientists are sometimes working on. New data that circumcised people are about 50 percent less prone to get AIDS is exciting, though whether this has to do with cleanliness or with immunological active cells in the circumcised foreskin remains to be seen. One thing we can be sure of is that circumcision is tradition, and not "our father in heaven" telling some people to become circumcised through the prophets he sends. If this "father in heaven" had the capacity to go into such details, why would he "intelligently design" humans with such a problematic prostate, or with an appendix, for that matter? Biology and evolution are, contrary to what many people tend to believe, imperfect, and they work by trial and error.

If you look at humans with compassion and enough distance in light of the anthropological data, it is not easy to say that we are a very intelligent species. But it is possible to say that we have been a successful

Bias Is Beautiful

species and that we are in all our inadequacies and biases a stubborn but affable one, ready to start a fight for any given reason and even readily kill each other as soon as we feel we can get away with marginal losses. But it is time to realize that we don't have much time on this earth if we keep going along in this manner. We have comforted ourselves with tales about trust and beliefs in our loneliness. But looking at the facts, we get the feeling that we must take our future (anthropologically speaking) and make decisions for ourselves if we want to have enough air to breathe and enough water to drink. To what extent circumcision is therapeutic will be resolved by medicine, despite ancient traditions.

Let us now go back to scientific matters that affect our everyday life a lot more. Alfred Kinsey, who was a biologist working on bees, noticed that to talk about sex was completely taboo, so he began to study human sexuality. He conducted widespread research, finding out that men masturbated, and published the results in the late 1940s. His report caused a revolution! Up until then, the cast of formal educators and clergymen had said that masturbation was a "sin." It was being practiced, of course, but because it was a taboo, everyone thought he was the only one masturbating, so the feeling of inferiority and sin was widespread. When Kinsey went on to say that women also masturbated, he landed in trouble with society. This was too much: Men did not want to accept that their mothers or wives (whom they had idolized) were also practicing masturbation. We cannot say that we approach the subject of sex naturally, even today. However, Kinsey has done humankind a great service by paving the path for people to talk more freely about these natural phenomena, which were even more of a mess of prejudices before his research. He initiated a paradigm shift in Western society, as all were masturbating anyway and burdened with a heavy conscience because of cultural bias.

I am sure that man will be able to find a solution for many types of cancer as well as for AIDS. That is, this will happen if we deal with first things first, be determined about issues concerning population, and act energetically! If we do not, then it will not matter whether we are Christians, Moslems, "black headed," "bald," "slanted eyed," or "black skinned"! I also do not believe that the population problem will be affected in any way by whether or not a man has been circumcised.

~47~

The Air We Breathe

Our real problems are simpler, such as the air we breathe and the water we drink! Air and water are essential for our survival: Freshwater sources comprise 2.5 percent of the world's water reserves and are used by half the world's population for luxuries like daily showers. There are people who walk twenty kilometers a day to find fresh drinking water. This statement is by no means pretentious humanism. Don't you get mad in the face of these realities? Aren't we the tribe that formed huge police organizations, armies, and insurance companies for our own security? Are we really so indifferent to the needs of more than half the world's population?

Money is important. It is possible to cut down a one hundred hectare forest and make a good profit. After bribing local administrations, paying out labor costs and transport fees, plus the sum spent for technical equipment, this operation will leave you with a sizeable profit. How about biological losses? Measured, for example, with carbon dioxide levels in the atmosphere. Think of forests as the earth's skin; how much skin do you need to live, or how much of your skin can you afford to lose and continue to live?

Don't you get angry if somebody accidentally hits your car or if your electricity supply is cut off for three days? Or if your Internet connection or telephone does not work for a week, don't you get organized with your neighbors and write petitions? Well, feelings and action for our environment are a lot more important than our cars.

There are captive air bubbles in icebergs at different depths, and it is possible to learn about the climate and atmosphere thousands of years ago from these samples. Just as we see the yearly circles formed on the cross-section of a tree trunk and can get information as to whether the season was dry or rainy in that particular year, scientists can evaluate thousand-year-old data from samples taken from icebergs. Without looking further into icebergs, let me talk about something else that we know today and can easily evaluate in the present time.

You now inhale with every breath six hundred times more chloride than Moses, George Washington, or Bismarck ever did. Humankind first started using these compounds about sixty years ago. These atoms do not have any effect on us as far as we know, but they do cause the ozone hole in the atmosphere to enlarge.

The ozone layer protects us from deadly ultraviolet light rays from the sun. Ultraviolet rays reach us through the ozone layer and help the Vitamin D production in our body, and therefore are useful. In high doses, however, ultraviolet rays can be strong enough to stop life.

Just as we have been able to live in good health in spite of doctors and without strife in spite of lawyers, and in relative peace in Western countries in spite of soldiers for the last sixty years, we should be able to solve these environmental problems in spite of politicians.

People need to breathe and drink water in order to survive. We know that through bacteria and plants, oxygen was produced on the earth that enabled the formation of higher organisms. Human beings, like other mammals, inhale this oxygen and exhale carbon dioxide, contrary to plants. When you light a fire in an industrial plant, this carbon dioxide is disposed into the atmosphere.

Let us look at the amount of carbon dioxide in the atmosphere, just as we looked at the population at the beginning of the book. Then, taking leverage from the stable parameters of science, let us move on to less stable domains of culture and cultural prejudices.

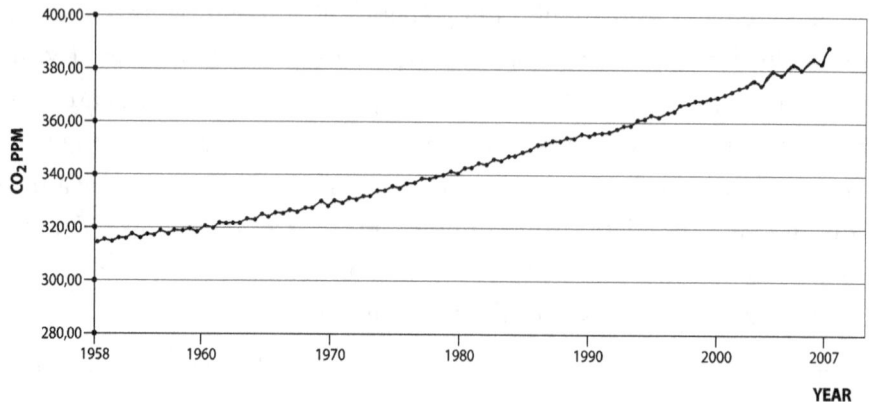

Fig. 4: 1958–2007 January and July CO_2 Measurements Made at Mauna Loa Observatory
(Source: Dr. Pieter Tans, NOAA/ESRL, www.cmdl.noaa.gov)

Forests and seas compensate for the buildup of atmospheric carbon dioxide, but both have unfortunately been pushed to the limit. These two cleansing forces have been fully exhausted. We came from the sea, and yet we are littering the sea with plastics and chemicals. I wonder how many of us take this to be some sentimental (pseudo) humanism. In fact, the amount of salt we have in us is the same as that in the seas. We have taken the seas inside us in order to survive on land and our kidneys strive to keep this percentage within strict limits.

Countries like America that are sensitive about these issues on the one hand have strict emission standards, and on the other hand continue selling cars that use leaded fuel to Asian countries that do not have similar laws. It is as if there is a deliberate break in our environmental consciousness, as if it makes a difference whether the damage we inflict on our planet comes from Asia or America.

The Kyoto Agreement, to which the United States is not a party, cannot be anything more than an experimental first step. In our world today, it is not possible to talk of a global solution if the United States is not an engaged part.

While reading the famous Stern report for the British government about global warming (published in the fall of 2006), I was truly moved, especially as someone who wrote a scientific article about air pollution and the health hazards it causes some thirty years ago. The Stern report is full of hope. One can almost see the fast pace, when economists take over from biologists. Speaking in terms of evolution, that's a very fast response. Allow me to get personal, dear reader. I am biased against social scientists, as I don't believe one can make a theory about anything if one has not worked in the biological realm for a number of years. This is my bias, or prejudice, or working hypothesis—call it what you will. I mentioned an article I wrote more than thirty years ago not to brag, but simply to share that such a track record gives me the courage to write about these topics. And to give the devil his due, once even social scientists (like economists) get their teeth into something, I believe things will start to change. This is the reason why I am so optimistic about this report. We are interest-guided creatures whether we like it or not, even if we often fool ourselves into thinking we may not be.

The Stern report briefly says:

All countries will be affected. The most vulnerable—the poorest countries and populations—will suffer earliest and most, even though they have contributed least to the causes of climate change. The costs of extreme weather, including floods, droughts and storms, are already rising, including for rich countries.

The costs of stabilizing the climate are significant but manageable; delay would be dangerous and much more costly.

The report also states that until 2035 a rise of two degrees in average temperature is to be expected if we don't get busy—a rise of five degrees in the average would be very dangerous indeed. This would be equivalent to the change in average temperatures from the last ice age to today.

I find it highly amusing that some scientists still publish articles doubting the effect of humans on global warming, or doubting the very fact of global warming. We are a stubborn and dumb species for sure. Well, I try to understand such scientists as crying out for help, saying, "I haven't grasped the big picture. Help me!"

~48~
The Ten Commandments Revisited

Our culture decides how we treat the environment. And lying and deceiving others and ourselves has been, and is, part of human biology.

Lying and stealing are also widespread among scientists. People simply think, "It is better to be famous than unknown," and that, "It is better to be rich than poor."

The struggle for wealth and fame in the world of science is just as widespread as anywhere else. We witnessed how the lies of a South Korean cloning scientist brought himself down, together with the hopes of many of his patients, in 2006. Another scandal broke out when it was found that in Norway a scientist who conducted research on cancer painkillers fabricated many of his cases, and what is more, other well-known scientists put their signatures on his papers. It is a known fact that scientists change their data to match the results they want, or plagiarize others' work. Just like businesspeople and politicians, scientists are not immune to such acts. Institutional lying can also be observed in the media.

Disinformation spread during wars is another case in point. I mentioned the example of Spanish flu, which during the First World War killed sixty million people. At first, people are silent because of matters of interest, and then it takes a while for such things to come to light. The weapons of mass destruction supposedly hidden in Iraq are one of the latest well-known institutionalized lies.

What is known as Piltdown man in anthropology (Eoanthropus Dawsoni, 1912) perplexed scientists until it was proven in 1953 that the skeleton was a hoax. To be an alpha animal is also a part of a person's impulses, and so is the desire to earn lots of money. As all is fair in love and war, by nature man lies when he thinks that his lies will not be found out.

Such examples are often found in daily papers. The point we keep coming back to is that humans are constantly in a state of strife with the rest of the world, and the more bias and conflict we have, the greater the probability that trigger-happy people will cause real mass destruction. There are just too many people on the earth, and the population must be reduced if our species does not want to become extinct.

When we consider anthropology and demography, we end up with human history and the history of cultures and religions. It is customary to use diplomatic language when it comes to religions. Only people like Bertrand Russell have declared their opinions boldly. Communism tried to avoid religion, but was obviously unable to fill in the psychological and social needs that religion helps to fill.

In August 2005, Pope Benedict XVI invited young people to Köln, Germany, saying that he would absolve them of their sins. Some four hundred thousand young people rushed there. According to press reports, their fare and bed and board were partly paid for by the church. People obviously want to guarantee a "happy ending" for what they do not comprehend and choose not to think about. They seek the easy way out. One can't help remembering that until Martin Luther's time, it was the custom for people to pay the church for absolution and not vice versa.

Prophets must have been people with high intellectual, social, and cultural intelligence who understood their times and could respond to people's needs. Their ideas about heaven and hell were not fraudulent; on the contrary, the prophets were people who were able to convey messages in a way the masses understood, thereby addressing the psychological needs of the people. Moses, who probably lived approximately twelve centuries before Christ, revealed the Ten Commandments. To Hammurabi's "eye for an eye and a tooth for a tooth," "to thine own self be true" was added, which seems to be the basic principle of all culture. Humanity has suffered greatly from people who were too biased to be honest enough to themselves about their ignorance and inexperience. Such people present themselves as possessing supernatural powers and thereby fool the ignorant masses (and possibly themselves).

Later, Christ added the dimension of love to religious beliefs. His "turn the other cheek" philosophy made people take into consideration others' points of view. "Father, forgive them, for they know not what they do" and compromise were other parts of Christ's teachings.

Some six hundred years after Christ, Muhammad added some rules: wash yourself, pray five times a day, and meditate with yoga-like postures while staying true to the same commandments. Most people know the meaning of the word "jihad" as "fighting for religion"; as I remember, however, Moslem thinkers have called it "small jihad." The "big" jihad is a person's struggle with one's self against egocentrism, laziness, hubris, and selfishness.

The last decisive step in forming our present culture is the understanding of evolution. I believe this was made possible in the philosophical and religious environment of England, a pragmatic kingdom with a high degree of religious tolerance. Although some religious people still have rather strong reservations about evolution, especially fundamentalists, Darwin's ideas have become part of our heritage.

When more people become educated enough to reach a consensus on democratic governments, the spiritual role of religions won't play a role in politics, as they do today. We can see all the elements of Western religions in all Eastern philosophies. Still, I would like to take another look at the Ten Commandments dispersed in the Holy Books of the Jews, Christians, and Moslems from the standpoint of the twenty-first century.

There should be no God but me! A monopolistic approach. No matter how many gods and goddesses one may have, the essence of this order should be tolerance and compromise. The separation of "them" and "us" was used as a reason for many a war.

Do not utter God's name in vain. What can I say? When a person is surprised and says, "Oh, my God!," this is a daily usage. However, it cannot be said that the linguistic daily usage of this term has prevented or caused wars.

Celebrate the Sabbath or the holiday. Working too hard spoils the psyche of even healthy people. A weekly pause of one or more days is healthy.

Respect thy elders, and love them. This matter is gaining importance with the aging population. Should we perhaps show more consideration to children than to adults? Tolerance and compromise are words that come to mind again.

Thou shall not kill! This should be observed indiscriminately. In Western cultures like Judaism, Christianity, and Islam, this refers only to humans. In Eastern cultures, it has a wider reference. And a subtitle saying "thou shall not torture" could also be added.

Do not commit adultery! Only physically? Or mentally as well? It is estimated that 10 to 30 percent of children born are not from the father on record, and now DNA studies are beginning to substantiate these figures in a variety of different cultures. An African proverb says, "Mother's baby, father's maybe." This, I suppose, is because people marry but want their children to have other genes they fancy more. Promiscuity is a part of our biology, and this won't change. It isn't a major issue as long as people don't kill each other over promiscuity issues. This concept has been broadened in scope in modern civil law as well. Fidelity is a matter between two individuals, and if these two are adults (both biologically and culturally), should not the choice be left up to them?

While the human male produces millions of spermatocytes daily, the human female (during the fertile phase in her life) often produces only one ovum per month. This basic biological difference is the reason for the difference of attitudes toward sex. The human female has evolved to be focused on pregnancy, which is a full-time job for nine months, and then must care for the baby for at least the next four years. "Caring for the baby" could mean anything up to twenty-five years in today's societies. The father's role biologically could be over in a few moments. This difference in commitment explains why the female attitude toward sex is concentrated on a lengthier period by nature, as sex is a central theme, whereas males' biological capability to focus could be over within a few moments—hence the male's genuine interest in the "goal" of a ball game. Ball games fascinate almost all men, whereas females are mostly interested in ball games just because men are. This basic difference between men and women and the extrapolations of it is yet another cause for hope for the future of humanity. The female biology is simply better equipped for processes of longer duration and for tasks needed to solve global problems. Females are a lot less aggressive than males and are involved about four times less often in violent acts.

Just as a forester is professionally "biased formed or deformed" to think in decades and a farmer in harvest seasons, the female is biologically built to have a longer period of focus than the male. This is a simple biological reality, which gives me hope for the future. Of course, all professional deformations or biases are cultural effects and when compared with biology, they have only superficial effect on our inner self.

U.S. psychologists Russell Clark and Elaine Hatfield sent an attractive female out to ask men for a one-night stand for the evening. Seventy-five percent of the men said "yes" on the spot; some of the remaining 25 percent excused themselves by stating a "prior engagement" for the evening. When an attractive male presented the same question to females, only 6 percent were ready to go to his place for the evening, and none promised to have sex.

As the population ages and the individual gains increased freedom, cultural frames of living together will surely change, and humans will look for new forms of coexistence. I hope this sentence won't motivate a busybody of a prosecutor to prosecute me for "propagating against the institution of marriage" for stating these simple biological facts.

The U.S. anthropologist Helen Fisher proposed a marriage license that expires after four years. Implementation of such a system could liven up the institution of marriage, as 97 percent of all mammals don't build teams of two for lengthier durations.

Cultures, hence also religions, limit our biological capabilities, and although they have affected and influenced us for hundreds or even thousands of years, from a social anthropological point of view, they describe only a short phase of our evolution.

Thou shalt not steal! I have doubts as to the applicability of this commandment. This impulse to steal is a form of competition for leadership. Stealing may also be undertaken to show that the thief is superior to the person whose goods are stolen. "I'm OK, you're not OK," and to show "one-upmanship."

Do not bear false witness! There is nothing to add. However, I believe we will have to fight to maintain this principle in centuries to come. This is an important commandment to observe for the spiritual welfare of the human race. The goal is to achieve the idea, "I'm OK, you're OK," but I guess it is still a utopia. Better technological control will help identify lying and false witness, I hope. And then:

Do not covet thy neighbor's wife, servant, ox, or ass! I hope the wife and the butler parts are overcome; the ox and ass can also be dealt with through trade.

We have a long way to go to mesh cultures with our biology.

I do not believe that stealing and lying will ever cease to be a part of human nature. The precautions can be preventive, but the punishment given will never be as forceful as the impulse to steal itself. Nevertheless, preventive measures may discourage stealing and lying from becoming cultural habits. Modern technology, such as electronic bracelets that confine "prisoners" to a limited, electronically monitored area, could ease the burden on jails.

Homicide is another primitive impulse. When will we realize that bombing people is an equally criminal act that should be brought to justice? In our history, only losers have been put on trial.

We can understand why Buddhism was popular with Western youths in the 1960s. This was not just a trend but a sincere search for truth.

Young people from Western societies were searching for the simplest and clearest truth in Buddhism and Eastern philosophies. The castes in Hinduism have also been surpassed in Buddhism. The five silas are don't kill, don't steal, don't lie, avoid sexual misconduct, and don't consume intoxicating agents.

The act of prostrating oneself while praying was integrated into Islam from Eastern (and probably from prehistoric) cultures, but prostration was also part of Mayan culture. The Christian cross and the swastika are also Mayan symbols. As there had been no contact between the Old World and the new for thousands of years, we can assume that these symbols of human culture existed at least ten to eleven thousand years ago, or five to eight thousand years before these religions were adopted. As underwater archaeology develops, we will probably find more signs of civilization in the Black Sea or what was once Mediterranean plains before the great floods.

Prehistoric cultures worshiped the sun and natural forces, but in the Ice Age they probably felt abandoned by nature. The first migrations to Australia and the Americas, as well as the later occurrence of Noah's Flood, may have been steps for humankind to initiate a belief in himself. Alone against the forces of nature, man probably began to worship an abstract ideal of his own self.

Just as the gradual change from hunting-gathering to a sedentary lifestyle was a step in human history mainly caused by demographic realities, the occurrence of monotheistic religions was also a landmark, with their beneficial and nonbeneficial aspects. Monotheism was most probably born from the Egyptian god Eton, or maybe even from the Sumerian An (meaning "sky" in Sumerian). Sumerians, who lived in today's Iraq, were probably the first culture to write down their beliefs. They called themselves "black heads," and their spoken language was agglutinative, Asian in character and different from other languages in the region. Once monotheism was there, Judaism, Christians, and Moslems followed suit.

Then the "Western" religions as we know them evolved, and as this is a book about demography, here are a few lines from Moses (Genesis 1):

So God created man in his own image,

in the image of God he created him;

male and female he created them.

God blessed them and said to them, "Be fruitful and increase in number; fill the earth and subdue it. Rule over the fish of the sea and the birds of the air and over every living creature that moves on the ground."

Because increasing population meant security against "beasts" and because procreating was something mankind liked doing and had to do for its own good anyway, these sentences were logical then. Looking back, we must accept the fact that religions did not initiate or participate effectively in major human achievements like abolition of slavery or women's rights.

According to the calculation of Fernando Savater, in a historical period covering fifty-five hundred years and encompassing approximately fourteen thousand wars, more than one billion people died, generally in wars that were struggles for power, but where religion was used as an excuse.

However, Savater's view is one sided and somewhat unfair to religion. While religion may generate intergroup hostility, the development of religious principles (so-called Judeo-Christian morality, for example) can also help to blunt such hostility. Within a particular religious group, such principles may also keep selfishness in check and foster community solidarity and cooperation. A further benefit is that many religions foster abstract, theoretical knowledge and reward those who acquire and share such knowledge, awarding them the title of "wise man" of some sort (monk, mullah, rabbi, and theologian). If divorced from day-to-day pressures and from ideas and principles one must simply accept or reject, this abstract cast of mind and respect for systematic observation and deliberate, formal education gives us the tools to think systematically about the future and may allow us the chance to shape it. And religions have given and will continue to give many "answers." Religions also have a soothing effect on people confronting uncomfortable questions like death and loneliness. Science has not been able to offer the comfort that belief systems have. So to make a long story short, I believe that matters about "god" are far too important to be left in the hands of men of the cloth. No matter what kind of uniforms they wear, every person counts and every person will have to make up his or her own mind about this subject. It is, in the last analysis, like health, which is also a very important issue for everyone, hence far too important to be left to people in medical professions only. Religions need to concentrate on new sins, like environmental ones, and incorporate them into their belief systems in the future. Religions trade with hope, and as the threat of at least worldly and physical punishment for "free-believers" is getting more and more unpopular, they will have to watch what they trade with and how they do so more closely, as trade with hope is a shaky affair.

Man is an egocentric and interest-guided creature. Like our ancestor the amoeba, we strive toward light and warmth instinctively. It is simply part of our survival instinct. We are a peaceful lot when fed and unpredictable and aggressive when hungry. Naturally, each creature is closest to his own self. When they realized that they were not alone, they put first the world (and later the sun) into the center of the universe. Then man realized that even the sun was not the center of the universe. Big Bang! What was there before that?

Although death is a practical cornerstone that shapes our lives, philosophically, I don't believe there is something like "death." Our molecules stay in the universe and our deeds and ideas become parts of future software, however minute, even if our species doesn't exist in the future. In other words, death is personalized bias becoming ubiquitory.

Here we come to the heart of the matter, its spirit: the concept of God, or as the Jesuit monk and paleontologist Pierre Teilhard de Chardin (1881–1955) called it, the "omega point" in "mesosphere." Should we perhaps call collective unconsciousness "mesosphere" in the light of new nanotechnology? Or call it "common human consciousness," as the thirteenth-century Anatolian mystic Yunus Emre did, or label it the memories of the Great Migration of the Aborigines? Famous Nobel

Prize winner and father of the Soviet hydrogen bomb, the late dissident Andrei Sakharov, wrote:

> For me God is: Not the ruler of the world, not the creator of the world nor its laws, but the guarantor of the meaning of existence—meaning despite all the apparent pointlessness.

The leader of the Duwamish Indian tribe, Chief Seattle, wrote to President Franklin Pierce in 1855:

> Every part of the earth is sacred to my people. Every shining pine needle, every sandy shore, every mist in the dark woods, every clearing and humming insect is holy in the memory and experience of my people . . . The white man . . . is a stranger who comes in the night and takes from the land whatever he needs. The earth is not his friend but his enemy . . . Continue to contaminate your bed, and you will one night suffocate in your own waste.

In short, we do not know as much as we think we do! Paul Watzlawick writes in his exceptional book *Munchausen's Pigtail, or Psychotherapy and "Reality" Essays* about a well-documented analysis of Wittgenstein's "ladder": "The world has neither a sense nor it does not."

We are simply on different steps on this stairway and each judges or perceives according to his "software" in a different way about what we can all see, feel, and read about. Not talking about it is no solution! On the contrary, talking about it is! Everyone will make his or her own conclusions according to the input he or she has had. Some will decide to pick up a gun and kill others. Some will look for solutions. Asking everyone to "shut up and be happy with the status quo!" is not really a solution, as everyone will work on his or her own conscience when given a chance according to the sum of his or her biology and experience (that is, derivatives of such things as IQ, strength, and amount of positive or negative feedback), though of course silence might be golden from time to time, if for no other reason than to give the other side a chance to reevaluate.

~49~

Lowest Common
Denominator

The lowest common denominator on which we can surely agree is that the concept of god should only be between "god" and the "individual." Such spiritual matters should not be the concern of others, especially not of any government. Ideally, the state should only consider the taxes it would get from the financial activities of religious organizations, and whether an individual's rights are really being respected in different religious groups. At least in the United States, all religious organizations are tax exempt—one reason, I suppose, why so many unscrupulous people want to be founders of new churches. Though we are living in an era where even some countries define their way of life with religions (and even enforce religious law), the identification of religious groups with particular signs and symbols and their alienation from other rival religious groups is probably going to become increasingly of historical interest as the ties of religion cease to be part of the state. The separation between church and state is ambiguous at best in any country: The ruling party in Germany is the Christian Democrat Union. The American president Bush often speaks of God, and in the United States a state-issued marriage license is required and priests can perform the actual ceremony.

In secular states, freedom of belief will become more equal and the worship of god will become more and more a personal way of searching for truth. The danger of one belief becoming dominant and thereby invading an individual's freedom will decrease in any society that can separate state ties to religion.

When we start cherishing our biases, then we start getting nearer to the solution. To regard prejudices as working hypotheses is a first step toward peace and constructive global environmental issues. To research scientifically, quantify, and measure prejudices are other steps that will eventually lead to solutions (solutions will not be found by remaining silent and absolutely not by hiding one's prejudices and cultural heritage within the family or among the members of a tribe, religion, or even a nation).

We need to be "selfish" enough to see this world as the one and only native country for all of us. We need to be selfish enough to understand that seeing the world as a single, united entity is necessary for our species to survive. This unifying attitude does not mean that prejudices will disappear. To have no prejudices is not possible and would mean total inactivity. Prejudice is a proposition and immediately creates an antithesis.

The evolution of our eyesight is responsible for the "blind spot," which every eye has. Man is hence unable to see his surroundings in an unbiased way. He is also unable to see himself without a mirror. Any person's appreciation of himself and his own history or his world is naturally biased. If a person doesn't do a task, even a very simple one, regularly, he won't maintain the perfection needed in an emergency. A simple example is the comparison of a clerk who runs one kilometer every day but works for a successful businessman who ran marathons on the national team forty years ago. If the boss has stopped running altogether and taken up overeating, alcohol, and cigarettes, who is the real athlete, the clerk or the boss? The one-kilometer man may think of himself as a lesser sportsman than the marathon man. In addition, the marathon man may think of the accountant as a lesser man altogether, but it is likely that the accountant will be fitter to run in an emergency than the businessman. This simple example can easily be applied to the understanding of successful crisis management in political or cultural problems. The example also demonstrates the need for objective and independent mediators in conflicts.

~50~
Meteors, Volcanoes, and History

The earliest impact that a volcano had on humans was probably the eruption in the Lake Toba region in Indonesia seventy-five thousand years ago. It was the largest eruption in the last two million years. According to geologists, the ashes in the atmosphere probably caused an arctic winter that lasted for years, resulting in the extinction of many species.

The Romans had swords made of a super steel called ferrum noricum produced from an iron ore found near Chiemsee in Southern Germany. According to some scientists, this ore was a decisive factor in the Roman Empire's domination, as their superior swords overcame the opposition in battle. Their steel swords gave the Romans a military superiority comparable to the guns the Conquistadores used to overcome the native population in South America. The quality of the Roman swords was only matched by the samurai at the end of the Middle Ages.

A meteor hit Chiemsee in 465 BCE, and the Celtic tribes came to the region a few centuries later and started making steel swords that were better than anything before. The Celts kept their metalworking technology secret and started selling these swords to the Romans. While examining honey samples from this area in the year 2002, scientists from Würzburg became aware of its peculiar carbon content. They thought the carbon itself was of presolar origin and had been brought to the earth by the meteor. They later found these same carbon atoms in the ore where the meteor had hit. Historians will have to rewrite Celtic, or Roman, or human history if the scientists agree on the presolar origin of these particles. This is the second "theory" I mention in this book that I believe is only a product of the originators' bias. People saying this probably forget about the Damascus swords and knives that dominated the weapon industry from the Iron Age to the Viking Age. Alexander the Great was said to have a Damascus sword, and even Aristotle commented on the high quality of the Damascus steel blade. Its origins can be traced back as far as 500 CE.

I personally don't believe in it, and many scientists don't either. The other theory I don't believe in is the "discovery of the Americas" by the Chinese Admiral mentioned earlier. I have mentioned these "theories" to show how "knowledge" develops. Humans by nature don't like the healthy option, which is simply "I dont know"! It is easier to have a theory to believe in. The Nordic steel and the better quality of Roman swords made from it was probably due to better furnace technology and better metalurgy than that of the tribes they fought. Their superior weapons probably had nothing to do with the meteor. But meteors did play an important role in the history of humans.

History records the year 1816 as a year "with no summer." In France, Switzerland, and Germany, beggars dragged around the streets in droves. In 1819, the first modern protests against Jews took place. Jews were richer and, of course, were a minority. Jews lived it ghettos (derived from the district Geto in Venice) in the Middle Ages, and during the Inquisition, many found shelter in the Ottoman Empire. The reason for the 1816 year with no summer, a bad harvest, and social unrest was the atmosphere. It is estimated that the lava of the Tambara Volcano in Indonesia caused the instant death of one hundred thousand persons in 1815. Clouds rose in the atmosphere and blocked the sun, the effects of which lasted for years to come. In New England, the whole summer of 1816 passed with snowfall and frost.

Human memory is short and is indexed to daily life. Eating, sex of some sort, entertainment, and then sleep is what we are usually most interested in. This scenario is our biology, and if we want to stay alive, everybody counts; his, her, or, in short, your opinion can make the decisive difference. Don't simply believe what you read here or anywhere . . . question the facts, and then get involved. Do research so that the contents of this book and others become a part of your daily cultural suitcase. You too can persuade others by singing, by telling jokes, or by sharing the facts you carry around with you. Everyone counts!

~51~
The French Revolution

If we go back to the French Revolution, some of us might even remember the names of its heroes. Words such as "equality" and "fraternity" were carved in our memories. Yes, we are all extremely equal as far as ultraviolet rays are concerned. You can be sure that this is true for dioxide and carbon dioxide, but what if that were not so? I do not think we are equal culturally, but human beings have the potential to be equal biologically, given time and opportunity. After a few decades or generations of comparable education and feeding, no substantial difference exists between people from different cultural backgrounds. An individual's breed does not matter, but if a person is lazy, even with the best education and nutrition, he will fail in every competition.

In 1782, the Hekla and Skapter Volcanoes in Iceland erupted. In 1783, the Asama Volcano in Japan spread lava and dust clouds into the atmosphere. At that time, Benjamin Franklin was in France and wrote the following:

> During the summer months of the year 1783, there was a constant fog in Europe and Northern America. This fog was of a permanent nature; it was dry and sun rays could not go through as they easily did through humid fog rising from water. These rays that could go through fogs were so weak that it was not possible to collect them with a lens and burn newspaper. Naturally, the heating abilities of the rays had extraordinarily weakened. As a result of this, the earth had almost frozen. Therefore the snow remained on the earth and did not melt and new snow kept falling.

Six consecutive years of bad crops followed. The winter of 1788/89 was very cold; hunger and despair were at their peak. The rest is in history books, explained with various interpretations. I find this simple "biological approach" to the French Revolution far easier to understand than the ones I had to read in history books. So much for equality.

Let's have a look at the present: Are all people equal?

You can easily see how three hundred million Arabs equal seven million Israelis. In conversations of this nature, reference is made to the Jews in the United States. Add them and all those around the world. Could you reach the figure of twenty million? Please do not tell me that I am a Jewish admirer; I have been dealing with biology (medicine, in other words) and agriculture long enough not to admire any group. In my experience, most Jews are ambitious and hard-working people (without forgetting, of course, that everything is relative). As individuals, Jews are obviously no different, but culturally, they differ from many because of their close "Mediterranean" family ties. When compared to people from northern cultures, they have been living with the urge for networking and have been working hard for centuries to produce a fruitful cultural environment of competition. Arabs, once the cultural center of the world, were dormant for many centuries and have had the curse of oil lately, which provided them with the income but not the need to develop the technology and the discipline to use it. I am sure with the humiliation they have had to endure in the last decades, very efficient Arabs will increasingly be noted on the world stage.

Wars are created from scarred bias. People who have been discriminated against for centuries have become the ones who discriminate. As mentioned, a Palestinian cannot become an Israeli citizen even if he marries a Jew. Israel has the nuclear bomb, and America has been covering it up at the United Nations. You can't measure fairness with different criteria for each group under consideration. The Arab-Israeli conflict must be solved for a more peaceful world, and I will write more about this necessity in the final part of the book, but let's get back to anthropology and history and environment first.

Through millions of years, our relatives and humankind were conditioned to have quick reactions. We needed to be alert to run away from a tiger or a snake, as we still need to be if we want to run away from a careless driver. Short and quick reactions were enough to survive. In time, humankind began making plans: plans to kill the mammoth, plans to survive the winter, plans for war, plans to cross the sea. These long-range plans were new abilities compared to the quick reactions needed for survival. The situation has changed in the last two hundred years; we are adjusting at an admirable pace in our own spaceship, and we are surely going to be able to make plans for the future.

~52~
Selfish Enough to Survive?

Will our selfish genes be selfish enough to shake us from our drowsiness and help us gather around a mutual proposition? I am sure this can happen.

What is the distribution of human intelligence? It is possible to understand with a single graphic: that of Gauss. Normally, distributions in biological colonies can also be understood with the help of Gauss's normal distribution function.

IQ and similar tests, no matter how much they are disputed, demonstrate a normal dispersion in each society. There are some differences between those who are right-handed and those who are left-handed, but in order to come up with an approximate figure, they can give an idea about intelligence distribution. If the top 25 percent and lowest 25 percent are set aside, one can find the 50 percent who have normal intelligence. Those with an IQ of 115 or above comprise the top 25 percent; approximately one-fifth of the top 25 percent, or 5 percent, are those with an IQ of 125 or above. By saying that there are exceptions to the rule, I believe that the majority of the top group is able to stand on their feet in any system. With their intelligence, they can, when threatened, form the habit of working hard and have the ability to adjust.

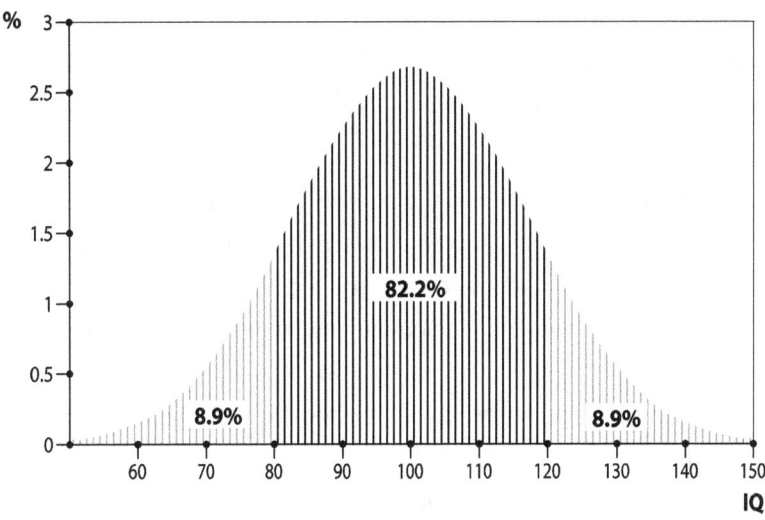

Fig. 5: IQ Distribution

The distribution curve isn't much different EQ wise. If we take the track record of what people have reached by the age of fifty (how many jobs they have helped to create, how much they have contributed to the common wealth), we will have a rough definition of the 10 percent of people who we need to work on a viable solution for us all.

Perhaps we should expect, through various tests, from the people in this vaguely defined top 10 percent, plans and developments for a hopeful future for mankind. If we wanted someone to represent us at the Olympics, we would look for people who could run fast; if we needed a solution to environmental and cultural problems, we should naturally seek people who have shown successful track records in problem solving.

It's not possible to know in advance which contribution to mankind will emerge from which race or cultural background. If our population numbers continue to increase, in light of scientific data, we will lose a great number of trained men through ridiculous conflicts and continue to poison the world. No two humans are equal, but for our own benefit, our cultural objectives should be equality of opportunity.

I do not believe that an IQ test can show all of a person's abilities and talents. A valid and reliable test to determine talent and ability does not (and probably will never) exist for all cultures. Even though IQ tests contain biases and are imperfect, they are useful instruments, which help, to some extent, to understand people within a particular society.

This 10 percent group mentioned above is not a group that can be determined through IQ tests. EQ and other differences within a population are other useful instruments that show the extent of human variability. The most valid test of all is success in the real world. Maybe we will have ratings in the future according to our records of accomplishment and the amount of community work one has done to be eligible to vote or have children?

When we look at income distribution, we see a similar distribution, with only a few people in the top percentiles. The richest two hundred people in the United States of America have more money than the combined sum of fifty million citizens of the same country. Worldwide, this curve is more noticeable; the richest four hundred people own more than the sum of the next two billion world citizens. At any rate, the top 20 percent, in spite of differences in each system, possesses more than 80 percent of the wealth of the world.

The first reaction of people to this reality is that it's not fair and has to change. As we see, there is no equality in biology. Rather than looking for fairness or unfairness in the reality I roughly pictured above by discussing only IQ and distribution of money, I think it would perhaps be more realistic to look for a solution within this framework.

In 1950, it was determined that 32 percent of the population lived in "developed" countries, and 68 percent in "underdeveloped" ones. At that time, the population was 2.5 billion. When it rose to six billion in 2000, the distribution became 80 percent underdeveloped and 20 percent developed. I would like to keep on rewriting those last two sentences over and over again, because they are essential for any realistic understanding of the difficult situation we now face.

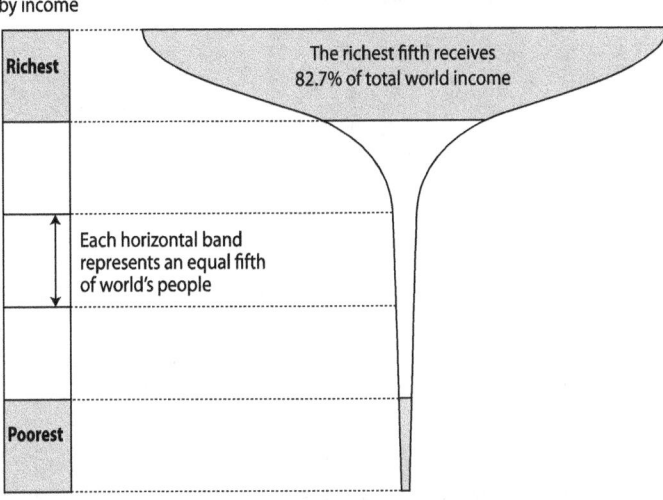

DISTRIBUTION OF INCOME

by income

Richest | The richest fifth receives 82.7% of total world income

Each horizontal band represents an equal fifth of world's people

Poorest

The poorest fifth receives 1.4% of total world income

Fig. 6: Distribution of Income. Wealth Distribution of the World; Indian Institute of Science

(Source: Mehta, Goverdhan, Science and Technology Society and the Knowledge Society; Indian Institute of Science; www.sciforum.hu/file/Mehta.ppt)

In years to come, the ratio for the developed is expected to range between 12 and 14 percent. There is no simpler way of explaining the reality that population growth makes people poorer.

When we look at the bureaucratic scandals in different political systems, perhaps it is biologically fairer for the money to be in the possession of people rather than for politicians or states to hold it. In general, people, especially if they have earned the money themselves, can be more flexible and more open-minded than those politicians who come into power with various ideologies and naturally perceive the world through ideological eyeglasses. They are also more open-minded than those bureaucrats who hesitate and are reluctant to make decisions.

Money is not everything, of course, and it would be more objective to evaluate a society's or a group's well-being within a society, maybe with a triad made of per capita income, rate of prisoners, and the longevity of that group rather than with a single aspect only.

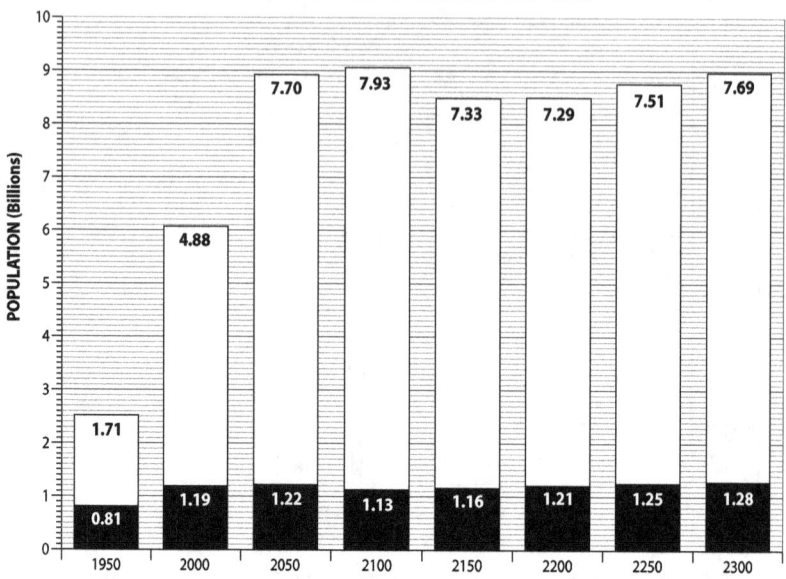

ESTIMATES and MEDIUM SCENARIO: 1950 - 2300

Fig. 7: Worldwide Distribution of Developed and Less Developed Areas (UN)

Even the dumbest human being realizes at some point that he cannot take his wealth along to the next world, in spite of all his efforts. Some wealthy people may build monuments as graves for their pets, but most wealthy humans are intelligent enough to realize that, as we know, even the largest of monuments are, well, just monuments. Their children, depending on their areas of interest, their intelligence, and their ambitions, can keep their inherited wealth or lose it. There is definitely a dynamic exchange in money. Because of longer life spans, people who do not have daily worries are taking active roles in various research and charitable foundations. It is hoped that this trend will continue in the centuries to come.

If we discount the probabilities of meteors colliding and if we hope that politicians will not force conditions that could bring about the use of nuclear bombs, we have to think that we will continue to live on the

Bias Is Beautiful

earth with only small disturbances. Perhaps it is an expected development that more people, with more knowledge, will be addressing world problems. Thanks to longevity (a unique situation in human evolution), more alpha animals, while they are still healthy, could approach issues such as Israel, Palestine, and Kashmir and make pragmatic and solution-aiming propositions.

In the years to come, I believe that new avenues will be opened for the distribution of justice. Think-tanks, with growing support from the public and media, look to the possibility of implementing revolutionary reforms in many countries.

It is estimated that in ancient times, the average lifespan was around fifteen years because of high infant-mortality rates. For around fifty thousand years, we estimate that this average age was decreased at times of war and during the ice age. This is a realistic assumption for the recorded date. My educated guess would be that the average age of humans was sixteen in the year 1800, rising gradually to eighteen years in 1900. The average age for the world estimated by the United Nations for the year 2000 was twenty-six. For 2100, they estimate it will be forty-four. It is difficult to foresee how effective gene technology and stem-cell treatments will be in the field of medicine, but the United Nations estimates that by 2300, the average age will rise to forty-six years.

Things will be changing in the near future in the world: According to the U.S. Census Bureau, the percentage of Asians and Hispanics will double, while that of the whites will go down by approximately 18 percent by 2050 in the United States. There will also be 35 percent more people over eighty-five in the United States than there are today.

The United Nations population estimate for 2050 is 8.9 billion, and we know from previous projections that demography is a reliable science. The projections made in the fifties for the year 2000 were spot on. It is projected that with the aging population, the rate of population increases on all continents will go down, and in 2100 will remain at nine billion and will continue at this level from then on. Just like what is happening

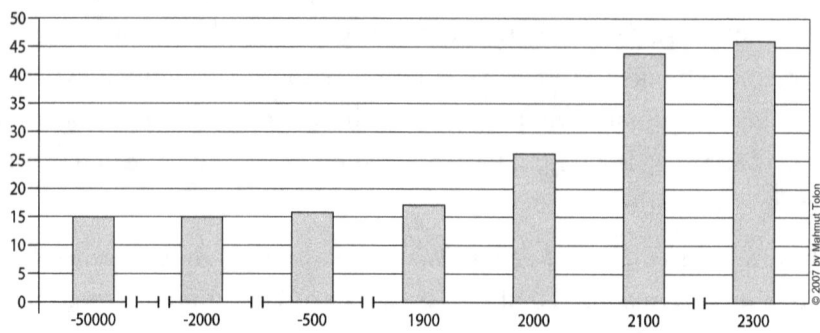

Fig. 8: Average age in the world in the last 50,000 years and extrapolated to 2300

in an aging anthill, humankind will have to live with the reality of a more crowded world. It is a fact that better educated people give birth to fewer children, but among less educated people, the birth rate is high. This is what demography is telling us.

Economically, China and India most probably will be the first among the countries with young populations that will close their balance sheets with positive results. In Western Europe: Germany, Italy, England, and in North America, the United States and Canada, with their aging populations, will set up new balances as far as production and retired populations are concerned. We should not be surprised if fifteen years from now China and India, countries that have made considerable investments in United States pension funds and in Wall Street, will reproach the United States, saying, "You have not given up your wasteful habits and are consuming too much of the environment." Ironically, the United States is doing exactly the same thing right now in China, referring to their human rights violations. This scenario, however, may not happen, as human nature makes us think that there will first be a consumption explosion in those countries as well.

The issue of aging populations will show its effects first in the countries referred to as "Western." The Western communities, which see the aging population partly as a load of ballast, will have to lead the world. Aging is a world problem, and thirty years from now, each country will have to face it. It is hard to believe, but in 2050, the number of sixty-five-year-olds in China alone will be equal to the number of all sixty-five-year-olds living in the whole world now.

What China has achieved in terms of birth control is simply fascinating. There are people who, with self-righteous pseudo humanism, criticize population control in China, saying that parents prefer male children and that sociological problems arise because the male population is now significantly larger than the female population. True, but is our choice simply for grown-up people to kill each other by strangling or bombing? Is our option to watch future millions starve, or is it to check the birth rate now? If that is the question, it is obvious what I will choose. Everyone counts: Every reader should decide whether or not this question exists and what his answer will be.

As long as our cultural biases remain as they are, I believe that the desire to have male children will continue to be significant in various cultures, at least for another half century. Because it is possible to tell the sex of a fetus by means of ultrasound, even without birth control, there will likely be an increase in the male population for some time.

Another stupid approach, in my opinion, is that of some small countries that say, "Our population density is not as high as in some other countries." The world is a biological whole, whether we want it to be or not. In addition, with our analytical ability and scientific intelligence, we are a bit more advanced than a simple "population divided by area" formula. It is necessary to set a more detailed biological ratio between the population and thousands of square kilometers in the Sahel Zone, or thousands of square kilometers of steppes and deserts in America, Eurasia, and Australia. In developed countries, when retirement pensions are calculated, a biological factor (it's called "genitive factor" in German literature) is also taken into account. For instance, in Germany, where the population is fast decreasing, the calculation of those who have children is naturally not the same as for those who are "egoistic," as they call it. After seeing the culture of two thousand years ago in

Ephesus, and after a better understanding of which mechanisms caused the collapse of various cultures, I am ashamed that we are still not able to reach a consensus and a minimal mutual understanding. Now we have all kinds of facilities to implement worldwide birth-control enforcement to regulate future population size.

For example, to keep our balance and live with the pleasure of doing our jobs, how many children from the East should correspond to one person in the decreasing population in the West, or how many trees should we plant to find the right balance? There are some such models, but they do not take enough space in our daily agendas. I am optimistic that population regulation models will be increasingly popular and will be known to the public and receive general acceptance.

We do not know when, or if, a climate catastrophe will occur, but it's certain that the aging population will cause a partial disaster, or maybe a revolution. How the future young population will carry the increased load of retirees will be left to a new generation. Any person reaching the age of thirty-five to forty realizes that he cannot run as fast as he used to. At age forty-five, he will realize he cannot see as well as he did, either. A war between generations is not probable, as young people realize that they too will eventually get old. However, there will be a pushing and shoving in the struggle for redistribution of wealth. The majority of voters will be elderly. For instance, in the United States of America, more than 70 percent of the capital will be in the hands of the baby-boomer generation born in the 1950s. A new word that has come into use in the United States is "ageism." We know that the "you are slower, less creative, in fact, you are a burden for me" approach has reached the extent where the self-confidence of the elderly is shaken. It is possible to see (at all times and in public places) such discrimination going beyond the borders of ill manners. Many elderly people, on the other hand, have made a habit of appearing in groups in the community that has hostile feelings for them (like the grey geese of Lorenz) in inconspicuous dresses of pale colors.

About half the people on earth are living in megacities already. This trend will grow as the population ages. Metacities with more than twenty million people are, and will be, our future. To name a few with more population than Denmark, Sweden, and Norway put together: Tokyo, Bombay, Delhi, Mexico City, Sao Paulo, New York, Dhaka, Seoul, Jakarta, Lagos, and Shanghai, followed by megacities like Rio de Janeiro, Istanbul, Moscow, Paris, and London. It is estimated that two-thirds of the world population will live in cities by 2050. These large conglomerations of voters with no experience in the pace of biology, largely people who sit in front of their computers and TVs, are, I believe, a serious topic we will need to focus on.

Common sense throughout the ages was held upright by the childbearing females who developed a realistic idea about biology and education. Maybe field work in one of the biological routines such as agriculture, patient care, or as a sabbatical to all civil servants, especially ones involved in international affairs, is to be discussed so that the majority does not lose touch with reality.

The advertising industry and filmmakers will need to redefine what they consider beautiful and dynamic images, which smell of youth fetishism. With the slowness in their activities and especially with decrease in their sight and hearing, the self-confidence of the elderly is being shaken even without the pecking of the young. But creativity and intelligence remains, and the capability of overseeing longer periods of time is something that one reaches through age. I remember that I could discuss the topic of population vigorously with people over one hundred years old. History is full of people who became creative after they were fifty: Michelangelo, Munch, Renoir, Goethe, Voltaire, Tolstoy, Verdi, and Richard Strauss, to name a few. People who can use their brains can do so, even when they are old, and those who cannot use their brains cannot do so even though they are young. The community can benefit from the elderly in the area of speeding up justice and in education.

Aging individuals often feel that they have "seen it all." Even the best movie can get boring with repeated viewing. In short, there is nobody who loves us more than we love ourselves, and we have to motivate ourselves and look forward to the future as humans. We need to overcome our simple biology and take responsibility if we want to survive.

We have to go toward a world where the number of newborns is in equilibrium with that of the dying. I hope we are selfish and intelligent enough to achieve this balance by using incentives to encourage new births after we find a healthy, steady state.

Human beings are equal because they have nothing else to lose other than their hearts, kidneys, and their lives. But they are not equal as far as their capacity to run, their ability for hard work, or their creativity or musical talent. And let me share a small secret: Humans do not want to be equal anyway.

Since the French Revolution, we have believed that all men are equal and that all pecking orders undermine human happiness. History shows, however, that not equality but equal rights and freedom to pursue whatever makes one happy without harming others is the best we have found, by trial and error.

A last word about equality: The assets of the world's three richest people exceed the combined GDP of the poorest forty-eight countries. Over one billion people live on less than one dollar a day and 2.8 billion on less than two dollars!

And a brief reminder as we talk about equality: In 1948, after the beginning of the Cold War with Russia, the United States's ally during the Second World War, the United States decided not to make a big "potato field out of Germany," as U.S. foreign minister Henry Morgenthau had suggested. The previous currency in defeated Germany was the Reichsmark, but also U.S. cigarettes. In June 1948, the Allied military administration in the Western occupation zones of Germany distributed a new currency that had been printed in the United States and brought into Germany in a secret U.S. military operation. The banknotes carried the name of deutsche mark: Each German citizen was given forty deutsche marks. This handout was the closest equality tried in history. Go to any orphanage and give each child ten U.S. dollars, then check it out after a week. Some children will have more, some nothing; let's wake up and enjoy reality; realities are always more fun than fantasies!

~53~
Saying Farewell
to the World

The day we die is the most expensive day of our lives. And most of this money is spent in the last months of life. In the United States, a cancer patient will spend thirty thousand dollars in the last year of his or her life, and 48 percent of this money is spent in the last months of his or her life. In modern cultures, we do not know how to face death.

Medical doctors are not well trained in this area. They have meaningful communication problems, whereas death is always with us and is a reality that we cannot change. When people are asked if they would want to know if they had a fatal disease, a vast majority of them will say "yes." There are unbelievable benefits for the community in doing research and organizing symposiums to discuss death. If people over the age of fifty were interviewed about euthanasia and what they would want if they had an incurable and debilitating disease, incredible amounts of money would be saved.

If you are curious about your future, you do not need to go to the latest Expo fair or to the space museum. Go to the nearest old folks' home and face the most realistic picture. Again, see the biological reality where human beings are equal. People you see there are elderly people—people just like you will be one day.

Euthanasia is still a taboo, but many stories can be found of nurses, doctors, and workers in old folks' homes who have killed nursing-home residents out of mercy. We read, register what we read, and yet somehow we do not really react. With conscious and inexpensive research, we will see that money can be spent more productively than for extending the lives of elderly citizens. While eleven cents of every dollar spent for medical needs is spent for stationery (red tape) expenses in Canada, this amount goes up to twenty-four cents in the United States. This difference amounts to one hundred billion dollars a year. These simple facts show that we do have a lot of room for discussion and reorganization.

When we treat community or corporate money as if it were our own money, we will be able to live more comfortably with lesser amounts without compromising quality. And it will cost less to die.

Technology is increasingly giving man the necessary opportunities "to control community funds." With life-span-increasing technologies, it will be possible in the near future to remotely measure blood pressure and pulse, to check fever, and to see if patients are mobile, in a very inexpensive way. With blood pressure or pulse data out of normal limits and with automatic alarms, it will be possible to determine where a person is, and these are technologies that will be widely used in the next decades (after, of course, reaching a consensus about the privacy issues).

There is reason to think that life expectancy will continue to increase several months each year in the forthcoming one hundred years. However, biology will eventually reach a limit. Leonard Hayflick observed in 1965 that cells dividing in the human lungs divided only a certain number of times and then died. The number of times they divided was always the same—a finite and limited lifespan. As the cells approached this limit, they started showing increasing signs of old age. This limit to the number of times a cell divides has been noted in all human (as well as in other) cell types; the limit is around fifty times. A fair amount of research is being done in this area, which could bring hope to cancer patients.

Besides modern methods such as gene treatment, diet, and physical activity, the usefulness of which we have known for thousands of years, it is possible to create meaningful differences in the life expectancy of the elderly and increase the quality of their later years.

The solution lies in our being selfish and in our cooperation to keep the world, our home and country, away from conflicts. Conflicts can kill educated people and cause environmental hazards any time for foolish reasons. The best hope for a conflict-free world is to control the world population so that it will not exceed seven billion.

After reaching a steady state, we should evaluate and discuss whether this world will be more comfortable with a population of seven billion or six billion or five billion, so that mankind will be able to continue to live on the earth. Some very serious and devoted scientists believe that mankind was in the best balance when the population was 2.5 billion (as was the case in 1950). It has even been alleged that the best ecological balance was attained with a population of only seven hundred million. These serious studies should be verified repeatedly. Constant measurements of the ozone hole and carbon dioxide in the atmosphere will become part of weather reports anyway.

Countries that are economic leaders can reach an agreement to limit the population to seven billion and then the United States, China, Russia, and India can take this proposition to the United Nations as a common project; can't this be a dream like Martin Luther King Jr. dreamed once and now realized? The solution is the agreement of all on the right of one offspring per person, two for a pair.

Each person should have the right to give birth to one child only and international support, such as rights and education, should be provided for that child.

Financial support for children exists in every culture as an ideal. Instances of one race "conquering" another race by increasing its population can be stopped by simple agreements, with each person having the right to one child. If we cannot achieve population control, we will see it anyway in the form of increased deaths through wars and epidemics, because it is not possible to fight with nature.

PART FOUR
BIAS AND SOLUTIONS

~54~
Prejudices

Even a single cell like the amoeba or the spermatocyte will strive toward success or survival. As a biological fact, interests will collide and even the single cell is selfish. Each more complicated being is an interest-guided entity and is hence naturally biased toward others, as there are inevitable conflicts of interest.

Imagine three blind people describing an elephant: The one that gets hold of the trunk will imagine the animal in one way, and the one who can touch a foot will imagine the animal in a different way from the first and from the one who can feel the ear. This example itself is biased toward people who can't see, and an evolutionary biologist specialized on the hearing sense will have different thoughts when he sees an elephant than a person who deals with leather or a butcher or a person who is courting an animal-rights activist. All will have different understandings of the "elephant," each being "deformed" or "formed" by his or her interests.

Our basic biological instinct for survival is common to all living beings. Every action is hence naturally defined by the will to survive (or the interests) of the individual, or the "self." The derivatives of our survival instinct are loosely named: the urge to multiply, attraction to the opposite sex, competition, selfishness, and the need for protection (houses, armies, insurance companies, banks, hospitals).

On the other hand, even the most self-centered, self-sufficient, and egoistic being realizes his or her helplessness and need for others. We need each other for survival and protection, and we have a strong herd instinct, which, in balance with selfishness, has been the cause of all our cultures and philanthropy. One should not underestimate this herd instinct: Loneliness and the need for a sense of belonging has defined groups, nations, and religions. The selfish instinct and its derivative the herd instinct are neither good nor bad but have influenced and formed our cultural and religious understanding of good and bad.

The relation between self and non-self is biased. In any community, the individual learns or has problems learning to compromise. The individual defines the group he is with as the outer self, but always in a more or less conscious monologue about "the others" within the group. The roots of jealousy, obstinacy, laziness, competition, pride, and insecurity are all within this monologue of the individual about interactions with the group. A constant accounting of positive and negative feedbacks defines the relationships. Ego and the way the individual can interpret feedback causes the multitude of different, unforeseeable actions and reactions. A Fortune 500 businessman once said, "If you want loyalty, your coworker must believe that you can think further for his good than he can himself." No relationship, even a marriage or parent-child relationship, is forever. If one of the biased parties believes that there is nothing more that he or she can take out of that relationship, the two parties will part ways. Once patterns are created and the sense of belonging and self-esteem is established within the group (family, social group, gang, party, company, or nation), the sense of identification may result in a stable relationship of years or decades—or in cultures and nations, hundreds of years.

The individual starts forming his neuronal migration as an embryo and in childhood according to negative and positive feedbacks, or lack thereof, and the experience and education that he or she receives as input. Each one of us becomes a unique entity. Each one of us is biased by nature. The struggle within the pecking order is the story of our lives. With communication and common sense developed through education, we can make the best of or even overcome this bias. This struggle, and the fact that we are all different and see and experience the world differently, defines the term "everyone counts."

Humans, like all living beings, are limited by their nature in perceiving the world they live in. Evolution has given us a "biased" or "imperfect" view of our surroundings (the blind spot in the eye, for example). Our perception of the world is interest-guided. The single human realizes how helpless and minute he is within his world and reaches out to other human beings to survive. This realization has caused the need for communication and differentiation, as we are selfish and biased and imperfect by nature. The struggle for a perfection that we never reach and for longevity and strength are parts of it. The common need for the others of our species has ended with a "common sense" that all humans share.

When setting out to define our basic bias as a new method for defining ourselves (or as a new paradigm, if you will), the English word "bias" is probably the expression with the least deviation of words like inclination and favoritism. So let's take this word to be the basic biological deviation we all have and the incapability of having a fair overview of ourselves and our world. Let's set common sense as the opposite of wisdom, which one can't have alone, and start trying to understand the world from this point of view. This is quite contrary to the hubris we have all developed, thinking we are the rulers of the world, bragging, exaggerating our achievements, either as persons or as cultures. We are all, one by one or as a whole, a dumb, obstinate, lazy, slow, and biased species, and this unglorified and realistic starting point in itself leads to the solution to a majority of our problems. Each of us needs others to survive. We also need to stop seeing bias as something negative. It is, in fact, something natural.

Cultural biases, or prejudices, are essential anchors in our minds that keep ideas from flying off in different directions. Groups could not compete effectively and civilizations would not develop without biases. They guide our thinking and hence our actions on both individual and group levels. Prejudice (Vorurteil) is the cultural word for the basic biological bias (selfishness and one-sided view). Prejudice is like a hypothesis, a conjecture based on our education and cultural background—simply put, it is the sum of each person's experiences.

We have seen the differentiation of self from non-self on the immuno-logical level. This also occurs on the cellular and social levels. As soon as we have an individual, a self, he or she will naturally differentiate himself or herself from others, from non-selves. As individuals, we constantly act in terms of self and non-self. We assess others to find out how useful they might be to us in terms of learning, work, entertainment, sex, and useful connections. If we see others as "not harmful, but useful," we will culturally register them as self. This also happens between groups of any kind and between nations. Everyone has both conscious and scarred (unconscious-historical) biases. Scarred biases can be destructive when embodied in political or religious systems. Biases can also be life-saving, as in evolution; species with "fear" of another can survive, and species without fear can be easy prey (usually to humans).

The desire to belong to a group is a basic instinct and a natural part of each personality. "Persona" is the Latin word for mask; in other words, we wear masks when in a group. We need the group for protection, self-esteem, and social interactions. But we also have to defend ourselves within the group. This mask, or individuality, however conscious or unconscious, more or less developed our individuality; it may be our sixth sense, if you wish. We do not always wear a "mask" consciously, but in the last analysis, it is in our drive (hardware) to be selfish.

We need to understand automatic biological processes like biases and pecking orders and to take responsibility for ourselves and for our world. No alien race will do this for us. The cultural belief systems we have developed so far have been a product of our laziness and self-glo-rification allowing us to believe in a supreme being that decides our futures. This bias has kept us from concentrating on the positive attitude (why worry if a supreme being will take care of it anyway?) and has given us hope. And this hope has given us the courage to destroy our surroundings, like vandals. Equating hope with fate and hopelessness with the readiness to work and think is another approach that would help us to tackle our problems.

Everybody operates on the basis of self-interest while equating self-interest with the collective interest of his or her own group. More people must understand that collective interest cannot be the goal of a nation or a belief system or a geography, but for the whole world.

No matter what race or culture a person comes from, any two men or any two women will eventually end up talking about "women" and "men," treating the sexes like two different "nations," or like self and non-self.

Here is another example: A man regards ball games, religion, or television as a waste of time and as the opiate of the masses (as Karl Marx might suggest); this is a prejudice. But this prejudice leaves this person with free time to do other things: play bridge, surf the Internet, or even work. This person will start classifying others as sports or TV addicts, while they will label him an elitist, workaholic, or computer addict. The amusing thing is that both sides actually follow the very same pattern and most bridge addicts are also football fans. We thrive on the typical division into self (which is good and the standard by which all things are judged) and non-self (which is bad, embodying characteristics neither we nor others "should" have). The people with whom we usually chat face-to-face are extensions of self, classified as "we." Others are extensions of non-self, or "they." But of course, whenever two people are talking about others, there is always a judgment of non-self about the other individual. If you don't say anything negative about others at all, you may be judged by your companion to be unimaginative. A certain level of openness is registered as warm and positive.

We are a biological product of our evolution. Culture, which we tend to overvalue (as we like bragging), is actually a sum of our prejudices (experience). There are no different cultures, only one. Culture changes over time. Herd instinct can be observed when birds migrate in flocks. On television, we watch the march of the penguins, birds that can't fly but can swim and walk. We perceive them as "nearer to self," as they walk on "two feet" in their tuxedo-like feather fur. Our common denominator with the birds is the pecking instinct, with higher- and lower-ranking beaks. When understood, this helps us see clearly the social impact it has on our everyday life. Nearer to us on the evolutionary scale are the mammals that usually hunt in packs. We also "hunt" in packs, as a family or in gangs or political parties or belief systems. The urge to migrate, which we find in all cultures, has already been mentioned: walkabout, tourism, and the weekend hunts of modern man for a bargain.

Every human being is defined by a multitude of neuronal migration: synapses caused by a multitude of experiences, cultures, and education. Each reader will obviously understand whatever fits into his or her own "biased" network, as is usual when reading the "biased thoughts" of any other fellow traveler.

1) People the world over have natural instincts to line up in a pecking order (if they feel it is in their interest) unless they see some injustice or harm done to what they understand as their own. The status quo is a good example of that. Many won't be happy with the present state and will try to influence the whole toward what they think is "better." The self/non-self distinction is part of our biological hardware. We can easily overcome most bias, which will make our communication a lot easier, when we understand others' biological background. Think about the bias of ageism. It is almost greater than bias caused by all religions, as it originates from the fear of death.

2) When people are given a clear line of thought that they believe is for a common good within a pecking order, they are willing to follow by nature, as we have seen in history over and over again. What they usually care about is what is good for them, and seldom what is good for all.

Religions have tried to emphasize the good of all out of need, as this "mutual need" became obvious in different phases of human evolution (for example, when we started asking questions about our very existence and about injustice in the world). This was the starting point for all the "Western" religions: Judaism, Christianity, and Islam. Many religions evolved, conveniently promising a happy ending, but they ended up discriminating against others. As religious ideas cannot survive in the stratosphere only, they were mixed up with ideas of power, and the self/non-self distinction occurred as a byproduct in conflict of interests. Communism, Nazism, and the Neocon Ideas of Huntington are examples of nonreligious lines of thought that people believed were worth following, but which have proven to be dead-end thoughts.

3) Wherever you have any sort of power, the rules are those of a pecking order. The people in power will help themselves first, and only then turn to others. In a democracy, leaders will be inclined to be populistic. When groups get larger, it is not possible for an individual to reach the politicians in power, because they will be busy considering their own interests and the interests of those who brought them to power. Over the course of history, we have seen that empires and religions became successful when they could survive with a relative minimum of bias. When they started having too much self-aggrandizement, they moved nearer to a fall. Kleptocracy is a natural byproduct of any larger organization. The better we all understand this tendency toward abuse of power, the less it will become. Democracy is an evolutionary stage that we have reached. It is better than anything that we have had before, but it needs adjustment. To overcome our problems, we have powerful, exciting new technology for communication (if we can share the ideas properly, that is, and agree on them).

4) The opposite of order is chaos, and we need order for a functioning system and security for production. Order and chaos are

like two ends of a flowing, dependant mathematical function. The more order, the less freedom, as order is organized by humans; and the less order and infinite freedom, the more self-ishness and insecurity and less ability to produce and plan and teach. With the recent population explosion in the last two centuries, we are fast becoming unable to teach what we have achieved to younger generations. Human education is a process that evolved naturally, copying or concentrating, cultivating, and magnifying what is in nature. It is also done naturally in a pecking order.

There is no "nation" in a racial sense of the world, only interest-guided herds, as biology has a tendency to mix. There is hope for the future in this and in the longevity of well-educated, successful people who have realized that they have made enough money or gained enough security and have freed themselves from daily chores for survival and realized the basics summed above. Longevity gives us the chance to work for the common good after the rat race. Never before have there been so many people of all ages who realize they have achieved what they want (or can get out of this world within normal limits) and who have been able to think of the world as a whole. Nor is there a single culture that defines all human heritage and meets the needs of all, as cultures are results of biological developments.

We need to find and define the cultural lowest common denominators and agree on them. We have self/non-self as a basic biological definition. What we have in differences between "cultures" is the heritage or wealth of human development. As illustrated earlier in this book, every living plant or animal can give a multitude of information that could be the solution or cure for a specific social ailment or medical illness. We need to tolerate each other as much as possible.

5) We can neither tolerate each other nor learn from each other if we (humans) feel threatened. And we feel "threatened" at a faster pace simply because of the size of the world's population. The distribution of

ideas and the ability to travel has increased in the last century, a very short period of time when considering the evolution of mankind. The only way to achieve tolerance is to freeze the population, strive toward a balance with nature, and constantly seek a new definition of this balance with new population sizes and investment in education.

~55~
Joy of Bias

Language, oft used as a weapon, is one of the most important tools for differentiating between self and non-self. The language of young people is different from that of older people. The multitude of languages in the world is easier to understand when you observe how language changes within a lifetime. Try to read a newspaper that is fifty years old. Also note that English spoken in New Zealand differs from that in Australia and that spoken in Texas and that in New England. It is the way groups define and recognize each other. Skin color has been and is still a sign of difference, as are possessions.

Speaking in a loud voice to someone who doesn't understand the language or inadvertently treating this individual like a fool is the most common method of intercultural (mis)communication. Raising the voice also increases blood pressure and stress, so such interaction may become physically dangerous.

Biases are generalizations that make life easier. Scientists or doctors, for example, can speak to each other in professional jargon that leaves the rest of us feeling left out.

But we need science and it is good that science is now being used by more people. Science in itself is searching for the truth instead of simply placing belief in something. Knowledge spreads faster and more educated eyes can make better observations. Scientific education allows us to observe things more objectively and systematically and to share what we observe with a minimum of bias. Science needs freedom of thought and time to develop and requires education. It is not as comforting as, for example, religion.

One danger of science is that it always needs critical evaluation, as scientists themselves are products of their own experiences and, hence, biases. Science is also not easily understood by the average person because of linguistic barriers that are partly because of educational differences and partly set up by scientists who want to separate and protect their interests from the public. For instance, doctors in every country make a point of using Latin words for technical terms to understand one another better, but also to mesmerize the public.

When something is made more complicated and more difficult to understand, we pay more attention to externals. We are more impressed by well-dressed scientists who exchange awards among themselves and appear on television than by "unsuccessful," largely invisible scientists without such awards. These less visible scientists may even be labeled "unreliable." The more complex and inaccessible science becomes, the more we fall back on symbols of prestige and influence, which may of course have little to do with the inherent quality of a scientist's ideas.

In his book on prejudice, the actor Peter Ustinov describes, in his entertaining style, prejudice as something "bad." However, I consider prejudice as something natural, sometimes even useful to simplify life. Often one is not at all conscious of prejudices. When biases remain unconscious, they can be dangerous. At the cultural level, this fosters what we may call "discrimination by othering."

The automatic, unconscious nature of bias (at least in the form of racial prejudice) can be seen in the studies of psychologist Mahzarin Banaji of Harvard University. She has clearly shown that while many individuals may honestly believe they are free of bias or prejudice toward members of non-self groups, neuroimaging and other biological and physiological measures show that this is not so. Neuroimaging can be briefly described as taking pictures of the brain with modern diagnostic equipment such as magnetic resonance imaging (MRI), flood flow, and "hot spots" while it is engaged in a particular task, like solving a math problem or sorting pictures into categories. Banaji's findings show that unintended, unconscious thoughts and feelings (biases, in other words) influence individual and group interactions. Unless made conscious and overcome, these biologically based processes make social justice difficult to achieve.

Violence can easily result when scarred biases are coupled with a lack of communication. On a group level, consider conflicts between skinheads and foreigners, and the lynchings of blacks in the U.S. South after the Civil War. Religious animosities and national rivalries are also good examples (conflicts between Shia and Sunni Moslems in Iraq, for instance).

Countries and leaders are also defined by their biases. The new leader of a country will enjoy freedom to live and act according to his biases. Biases will result in discrimination against people who do not support the leader. If the leader prefers his subordinates to be deaf and dumb, the communication gap between leader and followers will get wider and wider. In a democratic system, if you are clever enough to present things vaguely, nobody will sense your biases right away.

We can all agree that a "leader" who consolidates his power will enjoy a certain amount of freedom to choose assistants according to his philosophy, and that new people will start profiting from the new leadership. Profit and power is actually what political parties are after. Usually, in any country, sets of biases define who is eligible for election and who is not.

Relations among human beings are formed on the basis of positive or negative feedback as rewards and punishments. All feedback can be perceived by the leader as aggression, depending on the leader's psychology, state of mind, and mood, and can be reacted to with punishment or avoidance.

Humans have a tendency to isolate themselves from information and opinions that don't fit their biases. This can create a potentially dangerous situation for all by isolating the leadership.

It is our most primitive expectation that if the leader is not respected, there is punishment. This is simple biology, and a pecking order arises in human society. This is illustrated by the submissive position dogs take when defeated in fight. A dog gets into a fight with another dog in this "pecking order," but at the end, when it submits with the aim of "protecting its own species," the winning dog tolerates the losing dog. If submission is not demonstrated in the presence of others, a conflict arises. And if the fight (war) gets out of control, it can end with death.

As noted before, the self/non-self distinction exists even on the molecular level. In fact, vaccines are based upon this principle. Whenever an alien object comes into the body, body systems try to expel it. Humans are interest-guided animals, so naturally we size up people we meet, according to our biases, as useful or not useful in terms of fun, work, and sex. We make these assessments with our five senses, with our experiences, and with our biases: What results is a magnetic attraction or repulsion (or in computer language, a sum of 0 or 1). It is the sum that makes the famous so-called first impression we have of others in our mind's eye.

A historical sum of all this feedback forms prejudices. Sometimes a critical person might have a chance to correct them. If they are unconscious (scarred biases), then only education or an encounter with the problem could help. Or if these are collective scars that a tribe has against another, only getting to know the other culture can change this.

"Scars" lying underneath family and group memories persist. This influences relations between Jews and Palestinians in Palestine, Moslems and Christians in Bosnia, and Catholics and Protestants in Northern Ireland. Christians discriminate against Moslems, Catholics against Protestants, and Orthodox Jews against "liberal" Jews. There are, in fact, no clear rights and wrongs, only webs of prejudices and scars. Sometimes communities cannot excise such prejudices without surgical assistance.

Between different groups, tolerance or compromise is the zone between different sets of opinions, beliefs, or perspectives that does not lead to violence. Historically, different sects of Christianity (with the exception of Ireland) learned to "tolerate" one another, as the alternative was a lose-lose situation for all. If we are able to understand differences, we usually end up resolving conflicts without violence. This is difficult when one group dominates the status quo and actively disparages and isolates the other.

We get our fundamental training and a good portion of values in the family. Cultures adapt to geographical settings and then, when people move from one such setting to another, they retain their initial values (biases), which may now be of no use at all. Because it is hottest around noon, in hot climates people take siestas in the middle of the day, then go back to work and eat their last meal late in the day (around ten or eleven at night). In colder climates, people don't take siestas, work from nine a.m. to five p.m., and eat their biggest meal around six p.m. People with the siesta bias then often see others as overly rigid workaholics. The others often see the siesta people as lazy layabouts. I live with and like my own prejudices (but not anyone else's, of course!). They make life easy for me; I can avoid wasting time on what I consider "nonsense." I believe, for example, that French, Japanese, and British cuisines cannot compare with Chinese and Turkish cuisines (and I am very stubborn about changing this bias).

Obstinate norms and behaviors can be seen most clearly when groups move from one climate zone to another. Consider the former British Empire. When people moved from Britain to India or South Africa, they not only brought cricket and other games with them, but inappropriate styles of dress (wearing ties and top hats) and attitudes toward work. This is captured nicely in Noel Coward's song "Mad Dogs and

Englishmen Go Out in the Midday Sun." As one line says, "The toughest Burmese bandit can hardly understand it."

One bias that people are usually unaware of concerns the amount of personal space needed to feel comfortable when interacting with a stranger. As Edward Hall points out in his book *The Silent Language*, the space needed is much greater among people from colder climates. The classic case of potentially lethal bias and misunderstanding is when a German is on a balcony chatting with an Arab. The Arab keeps moving closer to the German, who keeps backing up until he falls off the balcony. Americans traditionally need more personal space than either Indians or Japanese.

Let me give some other examples of self/non-self thinking. Swedes put all Mediterranean people into the same category of "black heads." Turkish people, when talking among themselves, use the term "Arab" to refer to all "negroes and Africans," without differentiating among them. Turks only single out "white Arabs" from others. These are Arabs with a white complexion who do not have curly hair. Brazilians, on the other hand, use the term "Turk" to refer to anyone from the Middle East. Luckily, it seems as if these imprecise and negative systems of categorization are losing their meaning and becoming "politically incorrect" as more and more people start living abroad.

When a man who has believed all his life that women are less intelligent than men learns that Silvia Arroyo Camejo, a seventeen-year-old girl, has written a textbook on quantum mechanics, he is astonished. This is quite a contrast to the definition of women in the Old or New Testament or in the Koran, is it not? The most macho male will also be astounded, after having run one hundred meters, one thousand meters, or fifteen hundred meters, when he compares his speed with the track records of women athletes in any country.

Naturally, there are meaningful differences between men and women and between people from different cultures and geographical settings.

We have to understand and accept the importance of individuality. Lumping together everyone in a particular group obscures variability, or diversity, within a group. Just as it is impossible to know from which grass a life-saving drug will be extracted, it is not possible to know which human being will have the drive, courage, intelligence, and ability to have a positive impact upon the world.

Bias can easily be observed in athletics. Jesse Owens, the black U.S. runner who won gold medals at the 1936 Olympic Games in Berlin, caused confusion in Hitler's Germany. "Whites" had been led to believe that "negroes" could not run as fast as whites in competition. Until the 1930s, scientific books everywhere were full of such biases.

Many people die with their biases intact. They spend little time identifying and trying to correct their biases. However, an example of how biases can change is seen in Bertrand Russell. In 1929, Russell wrote the following:

> In extreme cases there can be little doubt of the superiority of one race to another. . . . It seems on the whole fair to regard Negroes as on the average inferior to white men, although for work in the tropics they are indispensable, so that their extermination (apart from questions of humanity) would be highly undesirable.

In 1952 he wrote the following:

> It is sometimes maintained that racial mixture is biologically undesirable. There is no evidence whatever for this view. Nor is there, apparently, any reason to think that Negroes are congenitally less intelligent than white people, but as to that it will be difficult to judge until they have equal scope and equally good social conditions.

Paradigms are sets of biases: Once the view changes, the whole picture becomes different and a serious change in the way we tackle problems can result. This is another reason for the general positive outlook in this book.

~56~
Dealing with Bias

I do not believe that a completely unbiased attitude is possible. An active mind is always full of biases. However, while a conscious bias can presumably be corrected when put to the test of reality, dealing with unconscious biases is more difficult. At the beginning of the book, we spoke about Wittgenstein's idea that if you can't say something, "then be silent." If everybody or even most people remain silent, people will overestimate the accuracy of their philosophy and come to believe that everyone else agrees with them. Psychologists call this pluralistic ignorance. So it is important to speak up. The more we speak, the more we will understand ourselves and one another.

In her book *Alles Getürkt*, Margeret Spohn beautifully demonstrates how inherited cultural biases are formed. In the Middle Ages, clergymen set up special anti-Moslem or "Turkish" prayers to warn the "public" against the ferocious Turks at a time when few people had the chance to travel. The effects of Sunday school brainwashing over centuries are shown in an objective way in this book.

In September 2006, Pope Benedict started a lecture by quoting the Byzantine king Manuel II (1350–1425):

> Show me just what Muhammad brought that was new, and there
> you will find things only evil and inhuman, such as his command
> to spread by the sword the faith he preached.

Why on earth did he feel the urge to use this quotation in the first place? He acted as his lifelong "culture" and "schooling" had taught him and was simply sending a message to people of his own flock. People do tend to forget what is good for their cause in their self-aggrandizement. But the pope really turned things over and was able to send a positive message to the world when he prayed in the Sultanahmet Mosque during his visit to the Orthodox Patriarch of "Constantinople."

Travel now is certainly much easier than before. Until the last century, people needed months or years to travel from one continent to another. In medieval times, only migrants, clergymen, and tradespeople traveled between cultures, carrying "naturally biased" information. The tradespeople were interested mainly in the money they could make, and the clergymen, naturally making their living through their belief, could only see different cultures as self and non-self.

If one cannot identify and frame bias, it is not possible to overcome it. If one lives with his or her own group (since "birds of a feather flock together"), one cannot overcome prejudices. This is something that each individual has to go through by himself or herself. Even journeys to other countries might not be enough. They often end up providing "data" to support preexisting biases, especially if the traveler does not know the local language and remains within his or her comfort zone. If there is a language barrier, people usually enhance their bias by staying in fancy hotels or all-inclusive holiday resorts.

Talking face to face in the same language or spending a few months together is the best way to overcome cultural bias. Because we haven't the time or resources to talk to everyone from different cultures or to understand everyone's language, television and movies are of great value and could enhance our understanding of different cultures, though to a certain extent they have helped to create a gap rather than understanding so far. Many in the Old World believe, for example, that all Americans drink as much as the people on the television series *Dallas*. And Nazi films did have an effect on the prejudices toward Germans.

A crucial aid in the reduction of bias is the biologically based desire to learn about others and to resolve conflicts to ensure our survival. It is upon this that we must build. Doing so requires increased awareness of biases and of interdependencies to enlarge zones of tolerance.

~57~
Inherited Prejudices

When a person has seen only two French families during his lifetime, he forms an opinion about the French based on such superficial contact. He may not realize that his views represent a gross generalization. He will, in any case, probably share his biases with his children without being aware of doing so. Prejudices about countries and races can easily be passed on from generation to generation in this manner.

When the son grows up and needs to make a decision (perhaps casting a vote), he will naturally act under the influence of his father's biases if he has no experience of his own. If the decision he must make doesn't force him to reflect and is as easy as pushing a button, he will react automatically. This may produce results that seem irrational, unthinking, and excessively vehement to an outside observer.

Political decisions, at least in "democracies," may be made in this way. A scientist raised in such an environment may act likewise when choosing from sets of statistics. The scientist will choose the statistics that best suit his own prejudices or biases.

If, on the other hand, the decision to be made is a major one that directly affects one's own life, the individual may do the research and engage in the reflection needed to overcome biases.

Businesses and advertising agencies try to create and capitalize on biases. This is what advertising and "product placement" in films are all about. After repeatedly viewing a product displayed in an attractive setting, you unconsciously come to prefer that brand.

Everyone has the following bias: that it is a waste of time to fight with bone-hard prejudices. The result and the danger of this bias is that it reinforces the tendency to mainly associate with people who agree with you.

We all live fairly happily with such harmless biases as "law is too important to be left in the hands of lawyers" and "medicine is too important to be left in the hands of doctors." Sometimes an event may disturb one's lethargy. The events of 9/11 and the invasion of Iraq that followed is such an event for me.

Anthony B. Robinson, a pastor of the United Church of Christ in the United States, wrote the following in an article of December 9, 2005:

> You might not expect a West Point graduate, Vietnam vet and career soldier to come out with a book titled *The New American Militarism: How Americans Are Addicted to War*. But that's what Andrew Bacevich, who now directs the program in International Relations at Boston University, has done.
>
> A self-described conservative, Bacevich argues that Americans have fallen prey to a "military metaphysic." By that he means all international problems are seen as military problems and the likelihood for finding a solution except through military means is discounted. The result is war as a permanent condition with the only acceptable plan for peace a loaded pistol. . . .
>
> How did this change, a crucial element of Americans' seduction by war, happen? Beginning in the seventies, a growing number of politically active religious conservatives told Americans, and their conservative Christian followers, that communism was everywhere on the march and America's subjugation was imminent. There was, however, not only this frightening side to their message but an urging to action. Christian America's true destiny is to wield military power in the death struggle with godless communism.
>
> Beneath this rhetoric lies a theology declared heretical in the early centuries of Christianity . . . After the Soviet Union imploded (in part due to its own military excesses), and 9/11 stunned Americans, these same politically active religious conservatives were quick to substitute Islam for communism. . . . A crusade

theory of warfare marched on, giving sanction to a new strata-
gem, "preventive war."

How and what happened may not be important from the anthropological point of view when we look at our evolution history. But it is of
utmost importance for all people living today. Americans used to say,
"What's good for General Motors is good for the country." In the
decades since that slogan originated, General Motors has turned into a
relatively unimportant industrial dinosaur. But the United States itself is
now more important than ever. We can't simply say, "What's good for
America is good for the world" and relax. We don't want America to
become history; it has produced too many good results. With the decision to invade Iraq, the United States set foot on an important but
treacherous path.

Pictures of Iraqi prisoners and their torturers recently went through
the world press, shocking many. Many felt pity for the tortured prisoners. That the U.S. soldier who joined the military to earn his livelihood
was sent to a faraway place and ended up torturing Iraqi prisoners is
understandable within the biased set-up of this war. So too are the accu-
satory and sometimes outlandish statements made by U.S. leaders and
Iranian officials. All of these people are doing what they think is right,
in line with their knowledge, customs, education, and experiences (that
is, their biases). When the subject is war, though, they may not really
"know" what they are doing in the biblical sense (Luke: "They know
not what they do").

Even voicing an opinion (when the going is rough) can be perceived as
a kind of aggression. To say "yes" to life is a matter of courage. So is
saying "no" to something convenient or something that one believes is
harmful. As members of the vertebrate family, let us courageously say
"no" to war and "yes" to securing our future.

The editor of the prestigious German daily *Frankfurter Allgemeinen
Zeitung*, Frank Schirrmacher, describes Samuel Huntington's book *The*

Clash of Civilizations as the "blueprint of U.S. foreign politics in the first Bush era." From all indications, he is completely right.

Samuel Huntington's book is completely wrapped up in a tangled web of prejudices that causes the author (and unfortunately others as well) to look at the world through the wrong window.

While written in academic language, Huntington's biases are clear. He identifies torn states as countries with a single predominant culture but whose leaders want to shift to another culture (such as Mexico and Australia). Core states are powerful and culturally central nations (such as Germany and the United States). Lone states, finally, are states like Japan that don't share their culture with any other country. Huntington tries to couch his understanding of Christianity (self) and Islam (non-self, what else?) in eloquent social scientific vocabulary. At base, though, he clearly feels threatened by the non-self hordes of Indios, Arabs, and Mexicans. In short, he fears that "his" United States is at stake.

Huntington, for example, worries about the future of English as a world language. He looks at the demographical distribution of languages and notes the tendencies in the different way different people speak English and says, "If at some point in the distant future China displaces the West as the dominant civilization in the world, English will give way to Mandarin as the world's *lingua franca*."

English will remain the major communication medium unless the United States continues to base its foreign policy on ideas such as Huntington's, alienating much of the rest of the world. All nations have made too many heavy investments in English for it to be dethroned as the leading language of the world, at least among the educated people who will provide most of the world's leaders. The differentiations of English spoken in different countries where it is the native tongue are simply because of self/non-self differentiation, which every country in its way naturally encourages. If one looks out of a similar window as Huntington does, every nation with different races or religions (most nations!) can be defined as heterogeneous or brittle, particularly the United States.

If we look back, as Huntington does, only about a thousand years, religion will seem very important. Looking further back, the importance of biology and the relative inconsequentiality of religion, or "race," becomes obvious.

Huntington's book, it seems to me, is the first time since *Mein Kampf* that a book by a writer who insists on looking through the wrong window is being taken so seriously. And I am not even sure that intelligent people took *Mein Kampf* seriously then. They may simply not have had the courage to speak up until it was too late.

Another theoretical dead end was communism. Those who took to heart what Marx and Engels wrote in the last century and founded new states on the basis of their understanding of their thesis had a chance to clearly understand that these theoreticians were wrong when they saw millions of people being killed in the name of constructing a new, more-just world order.

Communists made a religion of Marx and Engels' ideas even though these ideas ignored biological truths. Biology teaches us that people are not equal and will never be equal, no matter how social, political, and economic life is arranged.

I don't believe that the United States is under risk of becoming a new "devil," as Venezuelan President Hugo Chavez suggested in his famous United Nations address. I believe the United States will play a major role in the "new age of understanding bias."

~58~
Monotheism, Secularism, Pluralism

There is no black and white in any country, only shades of gray. But there is always a melody in this dance of grays. Allow me to tell you a bit about a "Muslim" country you might have heard of: Turkey.

In 1993, I wrote an article for *Milliyet*, one of the main newspapers in Turkey, examining the medical effects of fasting during Ramadan and comparing fasting in different countries. I pointed out that when Ramadan occurs in summer, fasting can produce serious dehydration that may be harmful to pregnant women or people with kidney disease or diabetes. I proposed that Ramadan should only be held in December, saving the lives of many and also contributing to the coexistence of cultures.

Shortly after my article appeared, I received a letter from Hezbollah (the Turkish Hezbollah was an entirely different organization than the one in Palestine; it was a local "club" of religious extremists). The letter began, "Hey, you infidel." It warned me that it was dangerous to talk about religious matters and threatened to slice off parts of my body. I talked with some friends from the media. Some showed empathy and suggested that I carry a gun. Others, some because they had not received such letters, gave me the impression that they thought I was trying to make myself important.

In 2000, we read in the newspapers that members of this organization had been caught and the group had been disbanded. Over one hundred corpses were found at the group's different headquarters, buried in cellars and gardens.

As a civilian in Turkey, it is hard to obtain a license to carry a gun, whereas prosecutors, retired police officials, members of parliament, and, of course, retired military officers are easily granted permission to carry guns! (Kleptocracy?) If a retired politician or military officer in a high enough place were to receive the type of threat I received, he would be assigned a bodyguard for life, paid for by the government. As the argument goes, these people protect the country from enemies and criminals and risk their lives, so carrying a gun is their "natural" right! The United States allows people to buy and carry guns, and that hasn't made for a less violent society. I am not a fan of guns, but as the reader might have noticed, neither am I a fan of elite bureaucrats.

The distribution of wealth is as important an issue in Turkey as anywhere else. Seventy percent of the people in large cities live in illegally built slums adjoining the cities. They are not yet "townspeople" with a happy place in the pecking order, nor are they anymore village people with some close contact to biology. The Turkish establishment is astonished when someone living in such housing goes to a bonesetter instead of an orthopedist, and even more astonished when this person becomes mayor of Istanbul and then prime minister. I refer, of course, to Recep Tayyip Erdogan. "Intellectual forces" complain about his manners and the fact that his relatives and close friends are fast becoming rich. For some time, the establishment had dragged this man from court to court for reciting a poem that referred to the resemblance between minarets and weapons!

On one hand, I am praising the upward mobility possible in Turkey (that is, "anyone can become prime minister"). On the other hand, I am chiding the "intellectual forces" for their complaints. The charming side of Turkish politics comes to some extent from the fact that "intellectuals" span the political spectrum: social democrats, conservatives, and bureaucrats (not unlike the situation in Brazil). Those on the right are more religious-oriented and populist, and are closer to the poor. At the beginning of the last century, Atatürk, the founder of the Republic, made significant reforms in his country concerning women's rights and the introduction of democracy. He was a general, and the Turkish military is traditionally very much involved in politics. The officers are well educated. The military keeps an eye on all social changes and operates

Bias Is Beautiful

almost like a modern university. This is the upside of the story and is followed by the inevitable downside. The officer is as professionally deformed and inflexible as any professional who has had any career.

A sergeant or a police officer, working hard "to protect the country," starts with three times the salary of a construction worker and has many additional allowances. Farmers and the majority of workers in cities are trying hard to have their children admitted into such occupations. What a surprise!

In Turkey, the military is such a deeply rooted institution that retired military personnel can drink a cup of tea at exclusive establishments, often located by the sea, by paying 20 percent of what a common citizen would have to pay. The military's retirement fund is also gigantic and very profitable. The military pension fund owns an important bank and is a major shareholder in the Renault car factory in Turkey. The officers live in ghettos and after a few years in the service the difference in perceiving realities becomes evident. The military has taken over the government three times in the last century in Turkey. As a result, three generations of scientists and politicians have been politically emasculated.

The other factor that influences political and everyday life is Islam. The Sunnis and Shiis are the main sects of Islam, like Catholics and Protestants in Christianity. Each has had a different development in each country. So the Sunnis of Iran and Iraq are very different. Alevites are the Anatolian version of Shiis and different in their everyday approach to religion than, say, the Iraqi Shiites.

If we consider the flexibility of humans and how many warlords have ravaged the country since ancient times, we have reason to believe that since Zarathustra, people of many different beliefs have adapted to the conquerors but kept their cultures and beliefs. But without any reliable statistics, almost everybody in Turkey is assumed to be of the Islamic faith. The figure of 99 percent is often quoted, but there is apparently a gag order of some sort about this figure. Tax money is used to pay the salaries of some eighty thousand (Sunni) imams. Roughly 20 percent of

Turks (though a recent Konda poll suggests only about 5 percent; the chief columnist T. Akyol suggested that the traditional shyness toward authorities could be the reason for such a discrepancy) who are Alevites don't receive any money from the state, because all Moslems are treated as Sunnis. The government does not provide imams to any mosque that calls itself an Alevi *cemhouse* (the equivalent of a Sunni mosque).

Calls to prayer are broadcast at high volume from loudspeakers on minarets throughout the country, without caring if there are patients who need rest and people who have just returned from night duty and are trying to sleep. Some columnists argue that loudspeakers are not part of Islam and that it would be healthier for the muezzins to climb the stairs of the minarets for the call to prayer. The government has forced its own brand of religion on the people since the founding of the Republic.

Government authorities apparently do not consider the public mature enough to completely separate religion from state business. I guess it is not stretching things too far when I compare the ideas of some Americans who would rather have "In Humanity We Trust" instead of "In God We Trust" on their currency. Even in the United States, with its strict legal separation of church and state, religion sneaks in, as we see in different speeches of President Bush. It is not much different in France, nor in Germany, where the leading political party is the Christian Democratic Union (CDU).

There are two sacred cows in Turkey. One is the "Atatürk" picture of the nationalists, who like to dot the landscape with his monument, and the other the "Islamic understanding" of the fundamentalists who want to impose their way of life on others. I believe the followers of both groups are no more than 10 percent of the population, but both groups are well organized (birds of a feather). I hope they will start seeing the world in a more biological way and start concentrating more on the inward values of their beliefs. If we remember that Mevlana, with his whirling Derwishes, lived in this geography over three hundred years before Shakespeare and thought with deep insight into human-nature tolerance, I am very optimistic. I postulate that as many as 90 percent of Turks would prefer a real separation of religion and state, that the government stop paying for mosques and clergymen (except for mosques of cultural and historical value), and that each citizen be permitted to contribute directly (and voluntarily) to a central fund that would help support all religious groups: Sunnis, Shi'a, Alevi, Russian Orthodox, Greek Orthodox, Armenian Orthodox, Armenian Catholic, and Jewish. Perhaps people should get some sort of tax relief for these donations. While mayor of Istanbul, the present Turkish prime minister said in a speech, "Islam is my reference." I hope his subsequent widespread travels around the world have broadened his views.

The Turkish public, weary of the slow judicial system, has increasingly begun to look to the European Court of Justice in Strassbourg, with some citizens winning cases against the Turkish government. In my opinion, bureaucratic and religious nepotism are the main reason the majority of Turkish people still want Turkey to join the European Union, as they believe Eurokleptocracy is better than the local bureaucracy. Among the youth, there is a serious reaction to the biased way the European Union is treating Turkey. I believe it will become a humorous footnote of bias in history if the "Christian culture" succeeds in keeping a country that has ruled one-third of the continent for about four hundred years out of the European Union.

The turmoil in Iraq after the U.S. invasion has demonstrated one thing clearly: It is not the controversy between the Moslems and Christians that causes trouble; it is human nature. The less educated people are, the more they behave as tribes. In Iraq, the main problem is between the Sunni and the Shii, which make up roughly 40 and 60 percent of the population, respectively. In one Moslem country, the Shiis are a minority (like in Syria) and in another they are a majority (like in Iran). If it is not a religion, it is a sect (like in Ireland). If not a sect, it will be hostility between Real Madrid and Glasgow Rangers aficionados. People tend to flock together when they strive toward power or happiness, and the more primitive and younger ones, especially the males, tend to fight each other. After a casualty or two, you start having scarred biases that hurt.

If a writer annoys the government, the government's reaction is often to sue him or her. The lawsuits turn into occupational therapy. Even if the writer wins, he gets sick of the judicial proceedings. In this way, he is "trained" to keep his opinions to himself. If the result is unfavorable, then off to the Sahel Zone! My optimism for Turkey is based on the fact that agricultural people and construction workers are still a majority in Turkey, and while many are religious, they in no way sympathize with fanatic fundamentalists. Even if a further shift toward a state with religious priorities were to occur in this country, it would not be a long-lasting one. The information travels too fast to satisfy the wishes of some fundamentalist militants who want to go back to the Middle Ages. The Islamic movement in itself was partly a result of world politics. Samuel Huntington's and other neoconservative theories of which the Iraq war is a result are pushing Turkey (and a lot of other countries) into the fundamentalist sphere. People tend to go back to their roots if they feel threatened.

Turkish newspapers have written about a female journalist who will be tried for committing the crime of "causing people to lose their good faith in the military." I can only say, "Oh my God!" I am sure that U.S., German, and Indian newspapers carry similar stories of events in their own countries. Of course she ought to be able to criticize the institution, even if most of the people revere it. The more developed a country, the faster the reaction of common sense to nonsense.

The following is an episode from the United States that I hope will entertain the reader:

Breaking News:

> In an attempt to thwart the worldwide spread of bird flu, American President George W. Bush has bombed the Canary Islands. Turkey is next. Intense negotiations with France about changing their national symbol from "the rooster" are being held.

I am an ardent online bridge player, and I told this joke during a bridge tourney I was directing last year on a U.S. Web site. The next day I found out that my tournament-directing privileges were removed by the owners of this bridge site, a site only devoted to the game of bridge. An amusing and rather harmless example of a nationalistic reaction?

We will willingly line up in terms of the prevailing pecking order and defend our alpha animals come rain or shine as long as the sum of 0-1 outcomes that the pecking order provides are positive. We will only try to change things if we are sure this won't harm us personally. As we embrace individualism and distance ourselves from tribal instincts, it will become easier to define and defend humanitarian values and differentiate them from pseudo humanism. Or, in other words, the more individualists there are, the more respect there will be for the value of self-interest. Humor has been a tool for dealing with the mighty long before court jesters, probably even long before people started writing.

The population explosion has worsened income distribution. Even more unhealthy, the age bomb is approaching, a bit later in the Eastern countries, but inevitably nevertheless. Jobs are scarce. These are universal problems with different densities in each country, and there are struggles over distribution of wealth. Doesn't this remind you of the analysis Michael Moore makes about the United States in his movies? Why do you think it takes years for the public to react once a country like the United States becomes involved in the affairs of other countries? Such involvement is mainly paid for with the blood of others, or, when American lives are lost, in the blood of the poor and of racial minorities. How interested the rulers of the United States or any other country are in the poor of their own country can be seen in the government's response to Hurricane Katrina. Let's be frank. Wouldn't the authorities have acted a lot more swiftly if the hurricane had hit Hilton Head or a prominent New England or California seaside resort? Cuba, with all its shortcomings, has fewer hurricane victims in a comparable geography simply because of better training, and a lot less traffic accidents because of better driver education. Western systems of democracies are not the high point of our evolution yet.

From the anthropological point of view, the idea of the nation-state is an ill-fitting dress. Nationalism and the idea that each nationality should have its own nation-state are ideas that developed during the eighteenth and nineteenth centuries and came to fruition with a dangerous vengeance at the Paris Peace Conference following World War I. With the help of new technologies and better communication, national boundaries may become superfluous, perhaps as soon as within the next century, as in Western Europe. And it has an opposing trend as well, of course, another side of the coin when we look at the Balkan countries that used to be joined and are now many small nations. Czechoslovakia is now two nations. The Basques and Kurds seek autonomy. Humans will always struggle forward through trial and error.

The "Western" monotheist religions (Judaism, Christianity, and Islam) have the tradition of striving toward power as well as wisdom, so it is a taboo of some kind for politicians in these countries to campaign without "religious baggage." Even in the most "liberal" of the three countries I have lived in, Germany, the Christian Democratic Union is the largest party. This "baggage" is larger in the United States and Turkey. The "Eastern" religions, such as Hinduism, Buddhism, or the Chinese beliefs, seem to be more concentrated in spiritual rather than worldly matters.

Most people don't have time to give secularism much thought because of the daily struggle for survival. They will strive for economic and social security and freedom of choice. Those who have more economic and political freedom are in a position to decide, to act, to express opinions. The choice these people make will reflect the sum of the individual's cultural biases, unconscious biases, and biases shaped by education, personal experience, and financial interests. These factors influence a leader's decisions and the decisions of nations.

But even if they have freedom of choice and action, people will be quiet and passive when they feel they can't understand seemingly complex political issues, so they just listen to the opinions of "experts" on TV about how many people may have been killed in military or terrorist actions. To them, this all seems like non-self! Even when they read about fellow countrymen being killed, they won't react, because these people are most likely from the non-self group within their own society. Only when a family member or someone they know is killed will they feel a threat to self. Steven Kinzer's book *Overthrow* illustrates how long it usually takes for Americans to understand and question U.S. actions overseas.

Communities and groups often act like an overweight person who keeps eating because he doesn't realize that he is full, or a rich person who keeps making money without realizing that he has earned enough for a lifetime. This stance can quickly turn into excess or overkill (literally and figuratively). If a culture doesn't get realistic feedback, it is likely to continue its preset patterns. This is true for any society. And how do we get realistic feedback if the communication we receive is disregarded because it is non-self? Do we need to kill off a friend or family member of each family to make individuals feel that they are personally involved? This would be a somewhat impractical approach, so we will need time. This book is nothing else but calling for birds of a feather to flock together for voicing what we need. What we need is to agree in practical and simple matters.

To make a decision for "saving the world," you seldom have enough proof. It gets its beauty from this tentativeness. You find new evidence every day and rearrange your understanding of this law of nature in your mind, just like the decisions each person has to make every day. Shall I trust others? Should I move forward and maybe speak up about the injustices I encounter? Or save my energy for yet more precautions? Sometimes a decision will be right and sometimes wrong. We have to decide every day, again and again. Or will we as humans be indecisive for a few more centuries and live on in set patterns like our ruminating relatives?

~59~

Germany

The two Germanies had a beautiful relationship, based upon tangled webs of prejudices on both sides, and both systems bombarded their "public" with their own beliefs: a country divided by a wall, one side "communist," the other "capitalist." We cannot say that the prejudices of Wessies in Western Germany and Ossies in the East ended when Germany was reunified. But the pace at which common sense was generated gives us hope for the future of mankind.

These prejudices were formed within the same country within one generation to such an extent that many people trying to pass from East to West Germany were killed as recently as fifteen years ago. The ideology was basically the protection of the interests of a minority: the ruling class of the communist bosses living in their *dachas*. The majority followed suit and started realizing that they didn't have the means to change. More and more tried to run away. The reason for illegal immigration is not very different. People will realize that they have one life to live and will strive toward light like moths. The creation of prejudices over a few years between citizens and people in power can happen in any country. Human nature readily creates groups based on self-interest and once in power wants to impose the rules fitting their own interest on others.

Germany was united within a few weeks. In 2006, one of the states in the Federal Republic of Germany started the plans to give a citizenship test to guest workers who wanted to become citizens. Journalists wrote that if this were obligatory for everyone, many people who were born in Germany would fail the exam. The situation is no different in America. To find a common minimum standard for who is eligible to vote, controversial as it might be, will be one of the tasks of any democracy in the future.

Germans had the opportunity to understand what happened in World War II as a result of an overdose of patriotism. This caused them to mature, and they are now ahead of other nations in their understanding of other cultures. (The Japanese haven't been so lucky.) There are of course skinheads and neo-Nazis in Germany, but most Germans disapprove of such groups. The late, postwar German president Gustav Heinemann once responded, when love of country or fatherland came up, "I love my wife; not Germany." A comment like this from a Mexican, French, U.S., Turkish, or Chinese politician would be enough to have him labeled a traitor.

Hitler was "democratically" elected. Most Germans preferred to act as Galileo did, bowing to power out of personal interest or fear. This was probably the right way to act in the short term, but very damaging for humanity.

Millions of lives were lost before the Nazis were defeated. And not only did Nazis pay, but all Germans, and future generations as well. For decades, Germans were given the roles of murderers in movies. In a way, even Palestinian refugees are paying for Nazi misdeeds.

One gets tired of hearing that the German culture produced Hitler. The people who brought him to power were simply a gang or a tribe or a flock of people who rallied around mutual signs and signals and ruled the majority. Most Germans at the time only worried about their own daily advantage and didn't dare speak up because they were afraid.

Let's look at it in terms of self/non-self and of a group of people herding together and becoming a group striving toward power: A few thousand people who agree on the basics of an ideology to seize power can make a huge difference. The docile rest of the flock will follow at first.

Is the situation any different in other countries? Was the seizure of power by the communists any different in Russia or Poland? Or the seizure of power by the Taliban in Afghanistan? What about Iran? Businesspeople who travel to Iran tell me that no more than 20 to 30 percent of Iranians like the way the country is run. On international flights from Iran, I am told that immediately upon take-off, many Iranian women uncover their heads and put on makeup.

Relations between countries even influence names for breeds of dogs. The Dogge or Great Dane or Danois was created in Germany. Only the Germans called the breed German (Deutsche Dogge). Everyone else called the breed "Danish." After World War II, however, the Germans themselves were so intimidated that they also started calling the dogs "Danish." With increased self-esteem among Germans, from the nineteen nineties on, they have resumed calling the breed Deutsche Dogge and have seen to it that encyclopedias also do so. When the French didn't support U.S. actions in Iraq, many will remember the suggestions that "French fries" should be renamed "Liberty fries" in the USA.

Freedom of thought is a process communities have to achieve themselves. Outside help can be of immense value, but is not decisive. Progress in a community is similar to the development of a child: You can pay for the child to attend the "best" schools, but many other factors will influence the child's development.

~60~
9/11

September eleventh was not a very important event for humanity from the anthropological point of view. It was only the result of prejudiced ideas of a mentally afflicted man more dangerous and biased than Charles Manson. Let's review how Bin Laden came to be.

While the Soviet Empire was an enemy, the United States provided arms money and organizational resources to groups of people neighboring the Iron Curtain countries where a majority of the population was Moslem. This was an understandable way to pursue U.S. interests. But a monster was created. The Taliban and Al Qaida developed in this atmosphere with U.S. support.

September eleventh was an insult to the West. The reaction to being attacked by a small terrorist group precipitated an occupation of Iraq (after clumsy stories about alleged weapons of mass destruction). Saddam Hussein could have been overthrown in less time with a more proactive approach without causing the deaths of thousands of Americans and possibly more than five hundred thousand Iraqis, and for a lot less money. The idea behind the U.S. invasion of Iraq was, in fact, to boost the economy at home. Noam Chomsky starts his book *Hegemony or Survival* with a quote from Thomas Jefferson:

> We believe no more in Bonaparte's fighting merely for the liberties of the seas, than in Great Britain's fighting for the liberties of mankind. The object is the same, to draw to themselves the power, the wealth, and the resources of other nations.

Chomsky relates how the majority of people all over the world sympathized with U.S. losses after 9/11 and demonstrates how, when the blind drive for vengeance took over, the United States acted as any spoiled, strong-willed child would act, feeling misunderstood and obstinate.

Did the American politicians who started the Iraq war realize they would be responsible for the lives of over half a million people when they were telling the world tales about weapons of mass destruction? No, I believe they were simply acting as any ambitious, self-centered, self-admiring, well-educated, and mostly monocultural powerful human beings usually act: to put it simply, dumbly. They ended up spending a lot more than they got out of this war. What we can all take out of this social-anthropological experiment is the wisdom that you cannot force people into things you believe to be true; you must work to persuade them.

It seems clear that the United States won't adopt the same approach with Iran. When watching the serious faces of Iranian leaders on television, I see humans, many of whom seem to have been a trifle malnourished during childhood. Most people in the region understand and sympathize with their arguments and their accusation that the United States is acting like a bully.

U.S. atomic submarines stationed around the world carry more explosive power than all the bombs exploded during the Second World War. There are also atomic bombs in Russia, France, Britain, China, Israel, India, Pakistan, and now in North Korea. Instead of initiating studies on international control of such bombs, the United States keeps making statements giving the impression that she thinks she will get somewhere by invading another country. This "I can have it but you can't" approach will generate serious confrontations, as has always been the case in human history. From now on, we must work to overcome this tendency. Verbal confrontations are taken as threats. The targets of such threats typically get frightened and engage in regressive or aggressive reactions.

When threatened, human beings fall back on what they have learned in the past. This naturally leads to a clinging to Christian customs and traditions in Christian communities, to Jewish customs and traditions in Jewish communities, and to Moslem traditions and customs in Moslem communities, until their fear dissipates and their curiosity awakens.

When human beings are worried about their own lives, they do not bother about cultural values and rush to survival. Both conditions are now present. We are all sitting on enough bombs to destroy the world several times over.

Many well-off parents in relatively poor countries send their children to school in the United States or Europe. Sometimes these students even receive fellowships or scholarships from their home countries, to which they seldom return. Immigration takes a somewhat similar form. Educated people from poor countries (particularly if they have money, as they often do) have a much easier time emigrating legally to better-off countries. Some even travel to the United States simply to have their children there so that their children automatically acquire U.S. citizenship. I am sure this phenomenon will capture the attention of the U.S. media soon and be included in the statistics of scientists a short while after.

According to United Nations statistics, there were eighty-two million illegal migrants worldwide in 1975, 175 million in 2000, and 191 million in 2005. In fact, migrants make up 3 percent of the world's population (more in Europe than in the United States) and no wall or fence will keep them out. We know that wealth distribution has worsened during this period and will keep getting worse if we don't react appropriately. Migration will keep increasing. While migration may benefit the migrants themselves, it may have a temporary negative impact (at least culturally) on the host societies. The more heterogeneous a society, the more conflict and prejudice there usually is and the more difficult it is to mobilize the population as a whole to successfully confront problems. Societies instead split up into mutually hostile, self-serving ethnically and racially based interest groups. As we once again see, there are always at least two sides to a problem.

Even if another absurdity like the Iraq war (scientists estimate some 650,000 deaths so far!) occurs in the Near East, like the United States invading Iran, or even if atomic bombs were to explode in the region, I think with the present cultural infrastructure this will set humanity back by twenty years at the most. Hence the basic optimism of this book. If an alien anthropologist were to look at the world, he would wonder at the stupidity of the human race for producing the Second World War, but also wonder how the human race was able to heal most of the scars of World War II and survive with so many odds stacked against it.

~61~
The Sum of Negative Outputs of Bias All Leading to a Positive End

We all live with a certain amount of bias, including about one's national identity. Unless forced by circumstances, many don't question this identity and some will probably be ready to die for it.

On the other hand, more and more people feel responsible for nature. Being selfish means nothing else but caring for the environment free of all boundaries and political systems.

Johannes Gutenberg's invention of the printing press brought cultural change. This occurred at essentially the same time as the discovery of America and the Ottoman conquest of Constantinople, because the need or the market for this technology became imminent. For faster, cheaper dissemination of written material to be important, there must be a significant number of potential readers. Our ancient ancestors knew how to reproduce written material with the help of woodblocks or clay discs, but there was no burning necessity because very few could read. Today we have the need and the means for communication to overcome cultural prejudices.

People's share of the wealth threatens to become smaller as the population grows, so people will continue to prey on each other, only more effectively thanks to modern technology.

Is it necessary to decide that population growth is to be prevented? Population is decreasing in more developed and in most better-educated communities anyway. This can be observed on every continent. Many better-educated people who care for the environment are deciding to have fewer children.

Active, intelligent, and alert people who are the most likely to sense the problems of the world need more wealth to feel secure, and they go after it. Their anxiety causes more inequality in income. In other words, the stronger will centralize when they are insecure and more greedy and this will cause yet more inequality: It is a vicious circle. We need a paradigm shift to break this circle. Once we can agree on what this shift is, it is easy to achieve. In the following paragraphs, you will see how simple and effective this shift can be. The inability of the better off to communicate with the poor leads to more anxiety and tensions within every nation. The same is true with the communication between rich and poor nations. As the rich usually are the more powerful and influential, the shortcut of wanting to export ideals such as democracy may look attractive. The reality behind this is, next to greed, the anxiety of the poor within their own country. Saudi Arabia was trying hard to export their own "Islamic approach" to all neighbors simply because the House of Saud felt they would be more secure at home in a more congenial neighborhood. Education needs time to mature and show effects, and not enough money is being spent on education worldwide.

Humans are obviously happier and more secure if they have jobs. With automation and technical aids, fewer and fewer people will be required to perform tasks. The rise in joblessness among young people foreshadows increasing social unrest. A supreme being will not come and solve these problems in the foreseeable future.

We have enough reliable statistics and knowledge and technology to aim for a total world population of no more than seven billion. With a population of five billion (perhaps an intermediate target?), quality can be emphasized over quantity. We can agree that we need balance with nature if Homo sapiens wants to survive on this world, but we have only started sensing this in the past few decades. Our species has transformed the world in less than a hundred thousand years to the point where our fertility has begun to threaten our survival. Remember that the dinosaurs lived for over a hundred million years and have ceased to exist.

The proposal leading toward a solution is simply to think over what many have thought about before: People with the capacity to give birth should acquire a license after more extensive training than that required for a driver's license. Each individual can have one child, which means that a couple can have two children. Every child born thus should be entitled to support from international organizations. People who have more children will incur financial burdens or penalties according to the laws of the country.

Restricting all couples to two children has distinct social advantages. Because the poor, as mentioned, normally have many more children than the rich, they will not pose as much of a political and economic challenge if the birthrate is equalized. We will have a chance for better education. Narrowing demographic gaps and being able to make accurate demographic predictions will also help to reduce conflict in areas with sharp ethnic divides, such as Israel and Northern Ireland. In Israel, for example, many Jews want to preserve Israel as a "Jewish state." About 20 percent of Israelis are of Arab descent, and even more Arabs might like to return to the homes they left in 1948, when the State of Israel was established. Because the Arab birth rate is much higher than the Jewish birth rate, there is a built-in conflict between the desire of many Israelis for both "democracy" and a Jewish state. If the birth rates of Arabs and Jews were equalized, Jews should feel less threatened and may thus be more flexible in solving the seemingly intractable and very bloody problem of Israeli-Palestinian relations. There would be similar positive effects elsewhere. In the United States, there are currently concerns about the balance between whites, African Americans, and Hispanics; in Turkey between Kurds and Turks; and in Germany between Germans and Turkish and other "guest" workers.

Although race, ethnicity, and social class are usually interrelated, a new form of ghetto is rapidly developing: the economic ghetto. Relatively affluent people now often deliberately wall themselves off from the less affluent, with whom they then often interact within rigid master-servant or employer-employee roles. This ensures poor communication across such divides and may not only diminish the desire of the well-off to benefit others, except in a defensive manner (to forestall revolution), but of the poor to view the rich as on some level "people," like themselves, from whom they can benefit in ways other than simply stealing their property. Whether there are walls and gates manned by armed guards (as in the Philippines and Indonesia), economic segregation is a growing problem in the United States, where real and imagined linguistic and cultural differences may magnify other differences between rich and poor, and zoning laws make mixed-land-use developments and developments containing different types of housing difficult. In Brazil or Canada, in contrast, the rich and poor often live side by side, while in the United States, "birds of a feather flock together."

With cultural globalization—the spread of television, movies, and consumer goods—it is impossible for individuals in non-Western societies (especially educated individuals) to feel that those in power in the world regard them as pagans and barbarians, if not worse. Although the United States may pay lip service to the fact that Moslems are just as good as Christians, many Moslems feel that they are considered "backward" people who, for their own good, cannot be left to control their own affairs. This is why the United States feels altruistic (and misunderstood) in its efforts to spread "democracy," even by force. One example of appalling lack of both cultural and psychological sensitivity and understanding was President Bush's surprise visit to the newly "elected" prime minister of Iraq in Baghdad in June 2006. The Iraqi prime minister was apparently informed of the U.S. president's arrival five minutes before the meeting, for "security reasons," according to Bush's spokesperson. But how could the prime minister and others in Iraq (even those supporting the United States) not feel humiliated by being treated in this way?

Ultimately of more importance, those most likely to feel humiliated by the West may be those from traditional societies who receive Western educations (along with philosophies of civil and human rights embedded in such education). They nevertheless feel looked down upon by the West and therefore identify more with the poor in their own countries rather than their Western would-be "mentors" and "benefactors." Virtually all leaders of anti-colonial movements, such as Ho Chi Minh in Vietnam, were trained in Western institutions implanted in their own countries. Closer to our own times, Osama Bin Laden is highly educated and comes from an extremely wealthy Saudi family. He could easily have gone into his father's construction business. Instead, he worked tirelessly against what he saw as disrespectful, dysfunctional Western influence in the Middle East. There is an obvious hypocrisy in the U.S. approach to Saudi Arabia. The United States speaks about democracy but supports the Saudi monarchy (obviously, because it wants oil and bases). Likewise, it presents Israel as an example of a democratic state while people in the area see it as a colonial enclave built upon land taken from Arabs (especially Moslems), with Western connivance. U.S. criticism of Iran for considering the development of nuclear weapons simply rubs salt into the wound. Why is it all right for the United States and such friends as India, Pakistan, and Israel to have nuclear weapons and not for Iran, which has been invaded and controlled by Britain and Russia and had its popularly elected prime minister overthrown by the United States in 1953 because Britain wanted to ensure its supply of oil from Iran?

In *The World Is Flat*, Thomas L. Friedman identifies two turning points in recent history: the collapse of the Berlin Wall and the development of the Internet. In my opinion, the last is as important to progress in the world as Columbus's discovery of the New World.

Friedman urges the United States to better educate and train its people, increasing the number of people with education at the doctorate level. I think this is not recommendable at all, not just for the United States, but for the whole world. I believe the basic race is run at an earlier age and we need to concentrate on kindergarten age rather than academics for grownups. Fancy a world with academicians (specialized in minute details of knowledge) in the majority, a world where we try to pursue common-sense talking with words like epistemology, dialectics, and semiotics. I don't believe that would contribute a lot to common sense.

I believe that living in another culture for a year or two and learning another language will affect our empathy for and understanding of other cultures. We have a long way to go yet. We have seen that humans vary in intellectual aptitude. It should, however, be possible to raise the ability to think critically and systematically by improving educational standards. But this won't lead to equality. Striving toward equality of opportunities is the most we can expect. People differ in terms of experience, courage, wit, sense of timing, and compassion. They will willingly group along their alpha animal lines. Maybe this plurality and the tolerance of it will bring about different ideas and the approaches we need to survive.

A doctorate or other advanced form of education has two different effects on us. The first is that the individual becomes a specialist in a well-defined area of knowledge. The second is that he or she better understands (hopefully) how much assistance he needs, learns toleration and cooperation, and recognizes how much effort it takes to share even simple insights in an unbiased way. On the other hand, the downside could be a false feeling of superiority and the expectance of getting more for this work.

Friedman provides some facts that are cause for optimism: One is that the software that allows the Internet to operate and supports two-thirds of the Web sites around the world (Apache) was created by several idealistic young men who had a favorable attitude toward IBM's offer to cooperate with them. They did not accept money from IBM but asked IBM to send its "best engineers" to work on Apache. The Apache people did not like some of the IBM engineers' work and sent them back. Friedman relates some other positive stories: the 100 percent increase in the funds allocated to fight malaria after the fifty million dollar donation made by the Bill and Linda Gates Foundation; the activities of investment banker George Soros; Jimmy Wales's establishment of Wikipedia, the online encyclopedia and dictionary, to share the world's knowledge for free.

Money and trade have facilitated the evolution of mankind. Humanism is an element in the essence of mankind. This drive to help is part of our herd instinct and is as old as human history. The people mentioned here are examples of this, no matter their race, religion, or nationality.

One can't help thinking and asking whether it is better for the wealth needed to solve the world's problems to be controlled by governments and bureaucrats or by successful people. The latter may be a more productive approach. From the biological perspective, it is high time that laws be made simpler and more just and that democracies be made simpler. Successful people need to guide us or at least moderate in solving international or national problems.

Perhaps it is to our benefit to turn our attention to successful elderly people who have run their races and are involved in charity efforts rather than bureaucrats and politicians. Most successful businessmen are not close to the public and don't wish to go into politics. But now, with a longer life span, we can benefit from their ideas and their creative energies, and we can encourage them to make suggestions for pragmatic solutions. Unlike bureaucrats, politicians, or professional scientists, successful retired businessmen who have created jobs for many are the real alpha animals who have become successful in the life laboratory. Unlike many a politician, most successful businessmen have satisfied their greed in the real jungle, and unlike many a professional scientist fleeing into small data and new expressions, they have proven their intelligence and risk-taking ability in real life, and they are generalists. As far as political arrangements are concerned, democracies are clearly superior to monarchies, dictatorships, oligarchies, and other forms of governance with limited participation by the governed. But they can be more productive.

~62~
Establishing Groups of Wise Men

In ancient times, humans needed the wisdom of the elderly, who were depositories of collective education and experience that they passed on to successive generations. Humans also relied upon the observations of those few (usually traders) who had access to other cultures, either through their own travels or contacts with members of far-flung commercial and religious networks. Once key languages assumed written form, some people wrote down and collected the experiences of travelers, adventurers, religious leaders, and military people who had contact with foreign cultures. Thousands of years later, libraries opened to a wider range of individuals.

Successful businessmen are more pragmatic, clever, and constructive, and are less interested in wars than many high-level politicians or uniformed or ununiformed civil servants. They have experience with money. Money is an important tool and other people's money has always been abstract. People work with others' money in a more careless and indifferent way than with their own.

It can be argued that politicians are doing the bidding of successful businessmen, even when going to war to protect those businessmen's interests. Active businessmen are alpha animals who live by the maxim of "all is fair in love and war." But when they realize that the race is run economically and that they still have fruitful years ahead, they could be enticed into offering solutions to serious problems we all face. Though I do have my doubts about whether they would have the flexibility and the tolerance to make good political leaders, I feel they could serve to make government spending more transparent, to make justice and education more efficient and available, and to work toward a new understanding of death. These people could define the new sins as the environmental ones and the penalties we need to agree on to survive.

What kind of people might serve as a present-day version of "wise men"?

1. People who have been successful in life as a result of their own efforts, rather than, for example, inheriting wealth, and have not worked as a bureaucrat for more than ten to fifteen years.

2. People with a good understanding of money and how to get things done, who have never been linked to or suspected of fraud or other illegal or immoral activities. Tony Soprano would not qualify.

3. People with no known ties to aggressive ethnic or racial or national interests, who believe in a proactive nonviolent approach both toward the state and individuals; people who have made enough money and have created jobs and ended their active business lives.

Between five and nine might be a good number for such a working group. Maybe later, five for each continent or each country. These people should be able to work actively at the solutions themselves, as not their money, but their capacities, which made them successful, are decisive for the task. "Ombudsmen" or "shadow cabinets" led by these could define approaches. Few would qualify for higher posts, but many could help out in the justice system or in creating better functioning welfare.

In the United States, Bill Gates and Warren Buffet come immediately to mind. They joined forces in the Bill and Linda Gates Foundation. We might all benefit if they were to look into the problems of this world and propose detailed, measurable solutions. These people would be charged with helping to solve problems in which they are not directly involved: justice and democracy are also fields in which we need active, unbiased contributions. Someone from Ireland and Canada, for example, might work on the Palestinian problem, and others from India and Japan on Ireland.

Eventually, the committee of wise people would acquire a broader base if such a group can provide constructive leadership. Members might serve for seven years each. Each country might come up with its own group of wise people and its own selection criteria. The starting point obviously would be people who are able to finance the offices needed and offer their services. Representatives from each national body might meet periodically in a world forum. Any pragmatic proposal made public could make a difference. Not only businesspersons but all retired people willing and able to work could contribute immensely to justice, health care, social welfare, and making bureaucracy efficient.

As a part of our nature, in every civilian organization, in each foundation and in every institution where people gather, there will be oligarchies inclined to take a monopolist stand in the pecking order. Utopias disregarding this pecking order will be tried again and again and will lead nowhere, as is the case in every political party and even in every marriage. When they find an opportunity or when they think there is an opportunity, human beings will be inclined to suppress the "inferior" and the environment as a whole. This is our nature. It's as simple as that! Now there is no other avenue we can take to protect our own interests, other than protecting our environment. The faster we realize this, the better.

We follow the contacts of the International Monetary Fund (IMF) with developing countries and those in need of economical aid, bureaucrats on one hand and politicians on the other. I am sure suggestions for quicker and longer-term solutions will come from successful and retired businessmen. Governments, even in Germany and in the United States, are incapable of distributing social aid in a transparent and fair manner.

~63~

New Signs of Hope

Communication is very fast now. More new information and knowledge has been generated in the last thirty years than the previous five thousand, and knowledge continues to double every three to four years, according to Goverthan Metha, director of Indian Institutes of Science.

President George Bush said in his State of the Union speech on January 31, 2006: "America is addicted to oil . . . The best way to break this addiction is through technology." In my opinion, this is a very important turning point. I feel these words showed the world that common sense is taking over from instincts and that the United States can show true leadership in caring for the environment and starting to adjust their lifestyle to environmental needs.

President Theodore Roosevelt said, "The settler and the pioneer have at bottom justice on their side; this great continent could not have been kept as nothing but a game preserve for squalid savages." U.S. anthropologists have since then brought to light the real history of the discovery of the Americas.

Martin Luther King Jr. shone as a leader in solving black-white discrimination, and the people of the United States rightfully began celebrating a holiday dedicated to his name. It is a holiday against prejudices, not one celebrated for a victory over another country.

Human beings are creatures that envy, cheat, and kill each other. If we follow the rationale of misunderstood biology that big fish eat small fish, there won't be anything left for us to eat, and other species will be victorious. Humans treat the world unfairly. Throughout human history, we have acted like grasshoppers in diminishing our own environment. With our improving technology, we have found methods of solving problems to the detriment of nature while comforting ourselves with the superficial belief that technology will solve everything. We are now realizing that heaven and hell are on this earth and that no one will come to solve our problems from up above. We are obliged to share in a more proactive manner the need for humanism arising from our core biology: the wish to survive.

The history of our increasing awareness of soil losses through erosion and our fight against it does not go back much further than one hundred years. The level of consciousness we have reached in such a short period is amazing. At the very least, there is an awakening feeling of obligation of those who understand to explain to those who cannot. When a target that receives a general acceptance is set, it is easy to achieve. We have to set the world population at no more than seven billion. To focus on this target will create new jobs and new motivations for bureaucrats and scientists.

Must population growth be prevented? Population is decreasing in more developed and in better-educated communities anyway. This can be observed on every continent. Many better-educated people who care for the environment are deciding on fewer children.

Active, intelligent, and alert people who are the most likely to sense the problems of the world need more wealth to feel secure, and they go after it. Their anxiety causes more inequality in income. In other words, the stronger will centralize when they are insecure and more greedy and this will cause yet more inequality: a vicious circle that only a new approach can break. As the rich usually are more powerful and influential, the desire to export their own ideals (such as democracy) may look attractive. As mentioned before, Saudi Arabia has been trying to export their way of life to all their neighbors, as the "Royal" Family will naturally feel safer if the flock around them does not get any ideas about democracy and such.

For better international or intergroup communication we need more and broader education. Education needs time to mature and show effects, but not enough money is being spent on education worldwide. There will continue to be fewer jobs for the uneducated as technology develops and automation takes over, and the rise in joblessness, especially among the young, foreshadows increasing social unrest.

Because a supreme being will not come and solve these problems in the foreseeable future, we must use our knowledge and technology to aim for a total world population of no more than seven billion. When 220 children per hundred pairs are born we know that demographically human populations reach a steady state. With two hundred or less children per one hundred pairs, the population would gradually decrease. With a population of five billion perhaps an immediate target over the next centuries, our species would be well on the way to taking our destiny into our own hands.

Then we can review the scientific parameters in the following decades to determine the ideal size population: the steady state that is fit for survival on this island.

Aiming at balance within the framework of data should shape the coming centuries. We have all the means to do this; the simplest way is giving a quota and educational support to each country. Just voicing this demand will have an effect in a short time. Organizations like the United Nations will be enlivened with a new duty, positive effects of which will be easily seen by everybody.

1. Each person's right to have one child should be protected; an education before having children should be obligatory. More than one child per person should be defined as a violation of others' rights and taxed accordingly. Obviously, the legal setup for this will vary according to the culture the legislation originates.

2. Freedom of travel. Living, working, or going to school in another culture for two years should be encouraged. I believe this will result in a far better allover understanding than, say, more "specialization," or more people with a Ph.D.

3. Continued commitment to education.

4. Improvement of the judicial system, preferably by involving retirees in areas outside of law, to continuously correct its mechanisms and to accept implementation of different legal principles and lifestyles in different communities.

5. Increasing oversight of government employees (public servants) regarding nepotism. This would increase everyone's commitment and motivation.

If we get used to looking at the world as a whole and if we agree on the basis of populations and the borders existing for the last twenty years, perhaps at the end of a certain period many borders will be lifted naturally. Militant and religious movements naturally cause biased deviations. We have to improve the tool called democracy. We are not a very clever species, and we have created our cultures partly under the influence of deceptions and evolution by flock instinct.

Lying to, stealing from, suppressing, and bullying subordinates by those in power will continue to be a part of cultures as long as humans exist. Our biological sins, like lying and stealing, will be easier to control with the help of new techniques and the availability of knowledge in the future.

Leadership is needed. Assigning a time limit to leadership is the best remedy we can find. The more we head toward laws less elaborate, simpler, and easier to comprehend, the easier and more comfortable our lives will be. When those who have religious and cultural convictions and those with none live together, the rights of the minority or the individual have to be protected. This is exactly what humanity has been aiming at from the beginning, with extraordinary successes in the last few hundred years. Yet another sign of hope is the uprising of the female sex. Biologically better equipped to understand lengthier processes and "liberated" in the last hundred years, more and more women are better educated and the trend is that they will have more political power in the coming centuries. Though the reality of today in many countries might make me sound like an incurable optimist, the trend is obvious.

~64~
De Gaulle and Adenauer

When we look closer at population control, we will find standing before us the ghosts of several thousand years. We have to do something and we have to do it better than our predecessors. We have to look for normal solutions without going to extremes.

This is what Konrad Adenauer and Charles De Gaulle, the leaders of Germany and France after World War II, did. They organized youth exchanges whereby thousands of youngsters from Germany and France lived and attended schools in the other country.

By increasing the number of such exchanges, by emphasizing biological similarities and by systematically emphasizing common cultural elements, many a problem could be tackled between unfriendly neighboring states. Issues between France and Germany, such as Alsace and Lorainne and two bloody world wars, were issues not easily resolved.

Another sign of hope was demonstrated to the whole world by the newly established State of Israel deciding to have Hebrew as its national language. Within a relatively short time (a few decades) a new language evolved. Many will say that Hebrew is an old language, because it has always been the language subscribers to the Jewish religion used in worship. It was not, however, an everyday, living language. Jews usually spoke either the language of the country in which they lived or such polyglot languages as Ladino, Yiddish, and Judeo-Arabic. A new, undated, everyday version of the language that could be used to facilitate communication among Jews who emigrated to Israel from many different countries was developed and inculcated within a few decades. It was like deciding and replacing Romanian, Spanish, and French with Latin in a few decades. Though the motive for this was a nationalistic approach, it clearly shows the ability to flock together around an idea believed to be of common interest in a short time. This is reason for optimism. Think of what could happen within a decade if a proper conception of bias could be instilled at schools and if people could agree on freezing the population?

It is a reality that developed countries damage the world's environment the most. What Jacques Cousteau said in 1991 is now truer than ever:

"The damage people cause to the planet is a function of demographics—it is equal to the degree of development. One American burdens the earth much more than twenty Bangladeshis."

If consumption in underdeveloped countries rises to that of developed countries, we will soon need more resources than we can possibly hope to find. Where do we find five atmospheres like ours? Developing countries are headed in precisely this direction. Today in Beijing, more than one thousand new cars are added to the traffic every day! Our survival instinct dictates the approach we must take. We cannot be so stupid as to continue damaging our environment. We must change, beginning with lifestyles in developed countries.

~65~
Killing Stares and
Instant Communication

Germans use the expression "if stares could kill." During a dispute, even a well-educated person could lose himself in a few seconds and, by reversing evolution, could reach a point where he wants to kill. With our sleep, sleeplessness, sex urge, or lack of desire, we are creatures who cannot do without eating two or three times a day. Just as we can aggravate ourselves into thinking of killing within a matter of seconds, so can we approach each other in a humble and constructive manner and achieve the impossible. Many statesmen have shown examples of this in history.

It is not possible to get anywhere with the approach "I have X, you don't, therefore I will bring X to you even if I have to torture you." "X" can be replaced with democracy, communism, Islam, Christianity, or Judaism.

Each cabinet or each shadow cabinet should include this approach in their mission statement. And each voter should demand it. Otherwise, if you threaten someone with a gun and if the only reason for his enduring your know-it-all behavior is your gun, you can be sure that he will find a way to get a gun for himself.

People who read and watch television saw the true nature of democracy during the elections between Bush and Gore. Although Gore received more votes than Bush, the electoral votes and Supreme Court decision favored Bush. In Florida, errors were found in the electronic voting. The media explained that some elderly people had cast wrong votes. A TV moderator said that elderly people were not as sharp as they used to be and in the news the picture of a demonstrator was shown carrying a simple slogan: "Stupid people shouldn't vote."

This is a glimpse of democracy and we have to accept it at face value. Press and TV are in the hands of richer and more active groups and such groups tend to be conservative and inclined to maintain the status quo. Democracy is a fairer system only when compared with absolutely authoritative kingdoms. When you remember that Hitler also came into power through elections, it can be seen that it is a system that needs to be worked on in order to protect individual rights. When a simple target is agreed upon, we can better define democracy and put in more checks and balances. Democracy is a phase that we have reached. It works with yes and no responses. And the politician tries to get more yeses. These yeses and nos are formed as a result of acquired knowledge, mostly of prejudices and partly of guidance.

In the evolution laboratory, a short gust of wind or a change in temperature creates the difference between "to be or not to be." Just as all higher species developed into beings with two eyes, a nose, and a mouth, we make choices every day to behave either like Galileo or Bruno and see only in hindsight that the majority of the choices we make are unnecessary and wrong.

Whenever injustice in distribution among flocks increases and causes disappointment for individuals, who sense diminishing hope for themselves; individuals' urge to oppose the system becomes more acute. "No future!" was the slogan of the youth in the seventies in Europe and is for billions now.

This is a function that can be described with mathematics. Someone who believes that he will not be able to make any difference either sulks and becomes depressive or, as was seen at Columbine and other disasters, puts his "after me, deluge" philosophy into action as part of a suicidal and lethal protest. On April 20, 1999, at the Columbine High School in Colorado, two young men killed thirteen of their school mates and wounded twenty-three others with semiautomatic weapons. A similar event took place shortly before Columbine in Port Author Tasmania: an armed man killed thirty-five innocent persons whose ages ranged from three to seventy-two. At the opening of the Berlin Train Station on May 27, 2006, a sixteen-year-old youngster stabbed twenty-six persons. The Virginia Tech massacre in 2007 was no different. I hope that the psychology and biochemistry of these aggressive and hopeless people will be understood more clearly in the years to come.

The romanticism of less than two hundred years ago becomes an increasingly realistic prediction, but we have the modern tools for a solution and a fair bit more knowledge now. William Wordsworth wrote:

> *The world is too much with us; late and soon,*
> *Getting and spending, we lay waste our powers;*
> *Little we see in Nature that is ours;*
> *We have given our hearts away, a sordid boon!*
> *This Sea that bares her bosom to the moon,*
> *The winds that will be howling at all hours,*
> *And are up-gathered now like sleeping flowers,*
> *For this, for everything, we are out of tune . . .*

When we describe and agree upon the target or "the enemy" or the challenge, our mechanisms will be more productive. What we need to realize is that there is no enemy, but that we have to take good care of our own ship to survive.

What we have seen was a biological glimpse of our history, a description of the human population movements, a definition of bias and the fact that an increasing number of the experienced elderly are in a position to look into world matters free of financial greed.

We are at the end of the book, and here is my proposal: a simple message, which will take about five minutes of your time and could change the world. Everyone should think about and come up with his or her own solutions. If we can reach a consensus and establish the needed organizational mechanisms, we could see positive changes within a few years.

Just put a simple sticker on your car or on your shirt, a "7" saying we should stop the world population at this number. Let's work together toward a better future. We have only one world. Moreover, if you question my convictions, my answer is simple: I believe in the superior race and in common sense! There is of course a superior and chosen race among us as defined by their biology. This is an open invitation for them. I hope they will be the first to put on their "7" T-shirts! To find out more about these superior beings here is a hint: Count the names listed in the acknowledgements on the next page and find out how many are male and how many female.

Acknowledgements

I would like to thank some people who have contributed significantly to this book. First, of course, my thanks to every reader who has generously given his or her time and has endured the bias demonstrated in this book.

We had a high school biology teacher in Bonn who took us bird watching in the nearby cemetery. As he usually had bad breath, we were all afraid he might come too close when explaining something. To him, next to my parents who were involved in agriculture, I owe my interest in biology. Our geography and English teacher was Dr. Paul Heinz Henke, who has been the president of the Dante Society (an Italian cultural society) for more than twenty years. He has since become a friend.

I want to mention some people who have further influenced me: In Kiel, I met Fritz Baade, an agricultural economist and colleague of my father. He had worked in Turkey and the United States during the Second World War (I believe that either he or his wife was partly of Jewish origin). Around his dinner table, we had discussions about the economy and environment. He was an adviser to German chancellor Willy Brandt and a leader of the Kiel Institute for the World Economy. Both Professor Baade and Professor Christiansen-Weniger, who lived on the family farm nearby, influenced my ideas about soil erosion and agriculture. Professor Christiansen-Weniger and his charming wife had lived for many years in Turkey and later in South America.

In Bonn, where I finished medical school, Professor Edgar Thofern, a specialist in hygiene and vaccinations, who was involved in the clean-up of the River Rhine, was my academic advisor. I am also indebted to Dr. Konrad Botzenhart, who contributed to my bacteria-centric world view. He later became a professor in Tübingen and a long-term friend. In medical school, I also met Professor Rupert Wilbrandt, a multilingual countryman of mine in the real sense of the word. During World War II, his father had worked as a professor of agriculture in Turkey, where he attended primary school and high school before studying in Bonn and in Cleveland, Ohio. A man with a deep insight into Jewish, Catholic, Turkish, German, and American cultures, he later became my partner in the Lithotripsy clinics (kidney-stone treatment centers) in Turkey.

While in medical school in Bonn, I met and later married Dr. Jutta Charlotte (Alsleben) Tolon. I am indebted to her for her contributions to this book. Our daughter Yonca Tolon-Sarigedik helped actively via the Internet, as a research assistant in New York. I also want to thank Professor Gül Güner, a biochemist. She not only helped with the research and translations but also in editing the manuscript.

My friend and colleague Dr. Juliane Schipulle (Germany) and her cousin in China, Mrs. Juliane von Hinüber–Jin, both went out of their way to help with the research for this book, as did Dr. Miriam Chernoff and Tonya Largy (in Boston) and Sheila Kassan-Michelman (Los Angeles), Ann Gogerty (Ames), Monika Buerger (Geneva), Patricia Landa Cragg (Lima), Professor Artun Ünsal, and Professor Umur Day Belge in Istanbul. I would also like to thank Professor Hayat Erkanal and his multinational team at the Limantepe excavation for sharing their knowledge with me.

The notes for this book were made in English, German, and Turkish. In detangling my notes, Erika Hess and Gül Güner were of great help. Dr. Stan Morse (Boston) also helped me, editing and with psychology. The bulk of the editing was done by Diane Özbal, and the final editing was done by James Gallagher. They all helped me make my notes more palatable to the reader.

No acknowledgements can be perfect. One will always end up forgetting a few people and situations: an idea overheard, a chat that left a mark years ago . . .

The principle is, as always, that x strives toward y, but never reaches y. This also applies to the literature for further reading. It would not be hard to provide page after page of references and suggestions for future reading. But I believe this would only impress the wrong people. I have therefore not cited literature for basic, generally accepted facts in medicine and biology and for other facts the reader can easily verify by typing the name of an author mentioned in the text into a search engine. You will not, for example, find publishing details for a book such as Darwin's *Origin of the Species*, or works of Konrad Lorenz, who studied the behavior of geese and was awarded the Nobel Prize after being influenced by Andersen's tales as a small boy, nor the *Textbook of Medicine (Cecil-Loeb)*.

Two books, which I read rather recently, have stimulated me to write this book. I start the further reading list below with these books and name the others as they come up in each section. Many have helped me, but none of them is responsible for my interpretations and mistakes. I enjoyed writing the book. I hope you have enjoyed reading it.

I saw the DVD of *An Inconvenient Truth* by Al Gore as this book was being proofread. For the first time, I was glad that he did not become the president of the United States. If he had become president, then he would probably not have had the time to make this movie. It is a movie that every world citizen should see at least twice. I hope it will become a truly global movie and make an immense contribution to our common sense and readiness to survive.

~Mahmut Tolon
Urla, Turkey,
June 20, 2007

Further Reading

Part One

1. Schirrmacher, Frank. *Das Methusalem-Komplott*. Blessing Karl Verlag: München, 2004.

2. Peterson, Peter G. *Gray Dawn: How the Coming Age Wave Will Transform America—and the World*. Random House: New York, 1999.

3. Gore, Al. *Earth in Balance: Ecology and the Human Spirit*. Penguin: New York, 1999.

4. Cincotta, R.P.J. Wisnewski, and R. Engelman. *Human Population in the Biodiversity Hotspots*. Nature (27 April): 2000; 404 990-992. http://www.populationaction.org/resources/publications/naturesplace/Nature

5. Klein, David R. *The Introduction, Increase, and Crash of Reindeer on St. Matthew Island*. Alaska Cooperative Wildlife Research Unit, University of Alaska. http://dieoff.org/page80.htm

6. Porter, Roy. *Blood & Guts: A Short History of Medicine*. W.W. Norton & Co.: New York, 2003.

7. Lyons, Albert S., Louis Lyons, and R. Joseph Petrucelli. *Medicine: An Illustrated History*. Harry N. Abrams, Inc.: New York, 1987.

8. Dychtwald, Ken, and Joe Flower. *Age Wave: How the Most Important Trend of Our Time Can Change Your Future*. Bantam: New York, 1990.

9. United Nations (UN): World Population Prospects, The 2002 Revision, 26 Feb 2003.

10. *United Nations: The Aging of the World's Population*. United Nations Gateway to Social Policy and Development, International Day of Older Persons, 1 Oct. 2004 (UN Program on Aging).

11. *World Population in 2300*. United Nations Expert Meeting on World Population in 2300. New York, 9 December 2003. www.un.org/esa/population/publications/longrange2/longrange2.htm

12. Natural history museums that can be visited online: British Natural History Museum: http://www.nhm.ac.uk/ Smithsonian Museum: http://www.nmai.si.edu/

13. DNA: http://www.pbs.org/wgbh/nova/sciencenow

14. DNA from the Beginning: http://www.pbs.org/wgbh/nova/sciencenow/

15. DNA: Prelude to the Symphony of Life: http://library.thinkquest.org/18617/

16. Ehrlich, Paul R. *Nufus Bombasi (The Population Bomb)*. Ballantine Books, New York. Ed. Dr. N. Tolon & M. Tolon Ayyldz. Mat. Ankara, 1976.

17. Tolon, Mahmut. *Keçi ve Zina (Goat and Adultery)*. SEL: Istanbul, 1993.

Part Two

18. Moorehead, Alan. *Darwin and the Beagle: Charles Darwin as Naturalist on the HMS Beagle Voyage*. TUBITAK: Istanbul, 1969.

19. Two DVDs about Darwin:
    Anonymous: *Evolution: Darwin's Dangerous Idea*.
    WGHB Boston Video: shop.wghb.org.

    Anonymous: *Charles Darwin: Evolution's Voice*.
    A&E.AETN.com. newvideo.com.

    The following Web sites are easily accessible and give a general overview of the journey of our species and demography of our cultures:
    https://www3.nationalgeographic.com/genographic/atlas.html
    http://www.bradshawfoundation.com/journey/

20. Cobb, Charles R. *Archaeology and the "Savage Slot": Displacement and Emplacement in the Modern World*. American Anthropologist, 4 Dec. 2005; 107: 563–574.

21. Kurlansky, Mark. *Salt: A World History*. Penguin: New York, 2003.

22. Eaton, S. Boyd, Marjorie Shostak, and Melvin Konner. *The Paleolithic Prescription: A Program of Diet and Exercise and a Design for Living*. Harper & Row Publishers: New York, 1988.

23. Lewin, Roger. *The Origin of Modern Humans*. W. H. Freeman & Co.: New York, 1992.

24. Redman, Charles L. *Human Impact on Ancient Environments*. The University of Arizona Press: 1999.

25. Festinger, Leon. *The Human Legacy*. Columbia University Press: New York, 1983.

26. Diamond, Jared. *Guns, Germs, and Steel: The Fates of Human Societies*. W.W. Norton & Co.: New York, London, 1999.

27. Diamond, Jared. *The Third Chimpanzee: The Evolution and Future of the Human Animal*. HarperCollins: New York, 1993.

28. Dawkins, Richard. *The Selfish Gene*. Oxford University Press: 1976 (2006).

29. Ehrlich, Paul R. *Human Natures: Genes, Cultures, and the Human Prospect*. Penguin Books: New York, 2002.

30. Morris, Desmond. *The Naked Ape: A Zoologist's Study of the Human Animal*. Random House: New York, 1969.

31. Ortayl, Ilber. *Osmanl Barisi (Pax Ottomana)*. Ufuk Kitap: Istanbul, 2003.

32. Pope, Hugh. *Sons of the Conquerors: The Rise of the Turkic World*. Overlook: Duckworth, New York, Woodstock, London, 2005.

33. Koestler, Arthur. *The Thirteenth Tribe*. Random House: New York, 1976.

34. Le Grand Dictionnaire Encyclopédique de la Langue Française. Editions L'Olympe, Paris, 1996. ISBN 2-7434-0797-2.

35. Der Kleine Ploetz, Hauptdaten der Weltgeschichte (World History), Verlag Ploetz, Würzburg, 1991. ISBN 3-87640-349-1.

36. Wikipedia: Die freie Enzyklopädie: http://de.wikipedia.org/wiki/Vladimir_I.

37. Harris, Thomas A. *I'm OK, You're OK*. Harper and Row: New York, 1967.

38. Martin, Paul S. *Twilight of Mammoths: Ice Age Extinctions and the Rewilding of America*. University of California Press: Los Angeles and Berkeley, Calif., 2005.

39. Perlman, D.L., and E.O. Wilson. *Conserving Earth's Biodiversity*. Island Press: New York, 2000.

40. Weiss, Ehud, Mordechai E. Kislev, and Anat Hartmann. *Anat: Autonomous Cultivation Before Domestication*. Science 16, June 2006; 312. www.sciencemag.org.

41. Kislev, Mordechai E., Anat Hartmann, and Ofer Bar-Yosef. *Early Domesticated Fig in the Jordan Valley*. Science 2, June 2006; 312: www.sciencemag.org.

42. Dubner, S.J., and S.D. Levitt. *Monkey Business*. June 5, 2005. http://www.nytimes.com/2005/06/05/magazine/05FREAK.html?ei=5089&en=6 bcb661222c32ba6&ex=1275624000&pagewanted=all.

43. The International Museum of the Horse: http://www.imh.org/imh/imhmain.html.

44. Lakoff, George, and Rafael E. Núñez. *Where Mathematics Comes From*. Basic Books, A Member of the Perseus Books Group, 2000. ISBN 0-465-03771-2.

Part Three

45. Le Grand Dictionnaire Encyclopédique de la Langue Française. Editions L'Olympe, Paris, 1996.

46. Horn, Christoph, Christof Rapp, eds. *Wörterbuch der Antiken Philosophie*. Beck'sche Reihe, München 2002. ISBN 3406476236.

47. Eckert, Roger. *Tierphysiologie*. Georg Thieme Verlag, Stuttgart, 2002. ISBN 3136640047.

48. Angel, J. Lawrence. *Health as a Crucial Factor in the Changes from Hunting to Developed Farming in the Eastern Mediterranean*, pp.51-68. *Paleopathology at the Origins of Agriculture*. By Mark Nathan Cohen and George J. Armeglagos. Academic Press, Inc.: Orlando, New York, 1984.

49. Jaquiss, Nigel. *Money Machine*. 6.07.2006, Naquiss at wweek.com. http://www.wweek.com/editorial/3231/7628

50. Tolon, M., et al. "A Report on Extracorporeal Shock Wave Lithotripsy Results on 1569 Cases in an Outpatient Clinic." *The Journal of Urology*. Vol. 145, 695-698, April 1991.

51. Hadler, Nortin M. *The Last Well Person: How to Stay Well Despite the Health Care System*. McGill-Queen's University Press: Quebec, 2004.

52. Bryce, Trevor. *The Kingdom of the Hittites*. Oxford University Press/Turkish Edition: Dost Kitabevi, Istanbul, 2007.

53. Dinçol, Belkis. "Hitit Yasalar› (Hittite Laws)" *National Geographic*, p. 84. Ocak, 2006 in Dinçol Ali Hititler, pp. 80–91, *National Geographic* (Turkey), Jan. 2006.

54. Savater, Fernando. "Die Zehn Gebote im 21 Jahrhundert" (Ten Commandments in the 21st Century). Verlag Klaus Wagenbach: Berlin, 2005.

55. Gorelik, G., with A. W. Bouis. *The World of Andrei Sakharov: A Russian Physicist's Path to Freedom*. Oxford University Press: 2005.

56. Dreyfuss, Robert. *Devil's Game: How the United States Helped Unleash Fundamentalist Islam*. Henry Holt and Co. (Metropolitan Books): New York, 2005.

57. National Academies Testimony - Climate Change Science and Research: Recent and Upcoming Studies from the National Academies: http://www7.nationalacademies.org/ocga/testimony/Global_Climate_Change_Policy_and_Budget_Review.asp

58. Heilprin, John. "Earth Is Hottest It's Been in 2,000 Years; National Academy of Sciences Says Humans to Blame for Global Warming." The Associated Press. http://news.yahoo.com/s/ap/20060623/ap_on_sc/global_warming

59. Schimany, Peter. *Die Alterung der Gesellschaft (The Aging of the Society)*. Campus Verlag: Frankfurt/New York, 2003.

60. Stökl, Günther. *Russische Geschichte von den Anfängen bis zur Gegenwart (Russian History) 5*. Erweiterte Auflage, Stuttgart: Kröner, 1990 (Kröners Taschenausgabe; Bd. 244).

61. Bundesfinanzministerium (German Ministry of Finance), BMF-Referat V B4 Leistungen der öffentlichen Hand auf dem Gebiet der Wiedergutmachung Stand: 31 December, 2005. http://www.bundesfinanzministerium.de/lang_de/DE/Service/Downloads/Abt__V/Le

62. Nadolny, Sten, Carl Honore, and Ralph Freedman. *The Discovery of Slowness*. Penguin: New York, 2005.

63. Mearsheimer, John J., and Walt M. Stephen. Israel Lobby and U.S. Foreign Policy. March 2006. www.lrb.co.uk or http://ksgnotes1.harvard.edu/Research/wpaper.nsf/rwp/RWP06-011

64. Goleman, Daniel. *Emotional Intelligence: Why It Can Matter More Than IQ*. Bantam Books: New York, 1995.

65. Fisher Helen Interview Der Spiegel Hamburg 9/2005, pp. 177–181.

66. ZDF: Terra X Der Cheimgau Komet. ZDF Expedition 08.01.2006. http://www.zdf.de/ZDFde/inhalt/22/0,1872,3262902,00.html

67. Vaulpel, James W., James Carey, and Christensen Kaare. *It's Never Too Late*. www.science.org. Science VOL 301 19. Sept. 2003.

68. Callahan, Daniel. "Death and the Research Imperative." *The New England Journal of Medicine*, Vol. 342, 654–656, March 2, 2000.

69. Mehta, Goverdhan. Science and Technology Society and the Knowledge Society; Indian Institute of Science PowerPoint Presentation. http://www.sciforum.hu/file/Mehta.ppt

70. Rich Spencer Health Care Paperwork. "Waste." *The Washington Post.* May 2, 1991.

Part Four

71. Devos, T., and M.R. Banaji. 2003. "Implicit Self and Identity." *Handbook of Self and Identity.* Eds. M. Leary and J. Tangney. The Guilford Press: New York, pp. 153-175.

72. Buchanan, M. "Are We Born Prejudiced?" *New Scientist,* 17 March 2007.

73. Tracey, Tokuhama-Espinosa, ed. *The Multilingual Mind: Issues Discussed By, For, and About People Living with Many Languages.* PRAEGER: Westport, Connecticut, 2003.

74. Kinzer, Stephen. *Overthrow: America's Century of Regime Change from Hawaii to Iraq.* Times Books, Henry Holt and Company: New York, 2006.

75. Ustinov, Peter. *Achtung Vorurteile (Attention Bias!).* Rowohlt Tachenbuch: Hamburg, 2005.

76. Russell, Bertrand. *Marriage and Morals,* pp. 266, 267. Liveright Publishing Corporation: New York, Horace Liveright, Inc., 1929.

77. Russell, Bertrand. *New Hopes for a Changing World,* pp. 113, 114. George Allen & Unwin, Ltd.: London, 1951.

78. Robinson, Anthony B. "Iraq War Debate Enters New Phase Debating the American Crusade." *Seattle Post Intelligencer.* http://seattlepi.nwsource.com/opinion/251384_tony09.htmlDec.09.2005

79. Huntington, Samuel P. *The Clash of Civilizations and the Remaking of World Order.* Touchstone Ed.: New York, 1997.

80. Chomsky, Noam. *Hegemony or Survival: America's Quest for Global Dominance.* Penguin Books: New York, 2004.

81. United Nations: UN Department of Economics and Social Affairs. Population Division: International Symposium on Int. Migration and Development. Turin, Italy, 2006; Population Aging, 2006. www.unpopulation.org (features e-mail subscription service)

82. Cousteau, Jacques-Yves. Interview. UNESCO Courier of November 1991.

83. Bush, G.W. State of the Union Speech. http://www.whitehouse.gov/stateoftheunion/2006/index.html

84. Burnham, Gilbert, et al. *Mortality in Iraq.* Lancet, 2007; 369: 103.

85. Akyol, T. *Türkler Kürtler ve Kalkınma ve Konda Milliyet.* 21.22.March 2007.